STO

ACPL ITEM
DISCARDED

3 1833 01664 3220

62
Calderone, Robert.
The complete
aviation/aerospace career

DO NOT REMOVE
CARDS FROM POCKET

ALLEN COUNTY PUBLIC LIBRARY

FORT WAYNE, INDIANA 46802

You may return this book to any agency, branch,
or bookmobile of the Allen County Public Library.

DEMCO

The Complete

AVIATION/AEROSPACE Career Guide

Robert Calderone

Allen County Public Library
Ft. Wayne, Indiana

FIRST EDITION
SECOND PRINTING

© 1989 by **TAB Aero Books**, an imprint of TAB Books.
TAB Books is a division of McGraw-Hill, Inc.

Printed in the United States of America. All rights reserved. The publisher takes no responsibility for the use of any of the materials or methods described in this book, nor for the products thereof.

Library of Congress Cataloging-in-Publication Data

Calderone, Robert.
The complete aviation/aerospace career guide / by Robert Calderone.
p. cm.
Includes index.
ISBN 0-8306-8280-5 ISBN 0-8306-8380-1 (pbk.)
1. Aeronautics—Vocational guidance—United States. 2. Aerospace industries—Vocational guidance—United States. I. Title.
TL561.C35 1989
629.1'023'73—dc20 89-31842
CIP

TAB Books offers software for sale. For information and a catalog, please contact TAB Software Department, Blue Ridge Summit, PA 17294-0850.

Acquisitions Editor: Jeff Worsinger
Technical Editor: Lisa A. Doyle
Production: Katherine G. Brown
Book Design: Larry McCoy and Lisa A. Doyle

Contents

Address Index

Acknowledgments

I would like to express my thanks and appreciation to the following organizations and individuals. Special thanks to the Boeing Company in Seattle, Washington for providing the job descriptions outlined in Chapter 3. The information provided was of great assistance and highly reflects the professional spirit that has made it one of America's finest organizations. Sincere thanks to the Federal Aviation Administration for providing much of the information outlined in Chapter 5 about jobs with the U.S. Government and in other areas. Most people do not realize how much information is readily available from the government at no direct cost. The following pamphlets were utilized: GA-300-122-84, GA-300-123-84, GA-300-124-84, GA-300-128-84, AC-140-2S, and AC-147-2Y. Thanks to Kit Darby and all the helpful people at FAPA (Future Aviation Professionals of America) for providing much of the pilot, flight attendant, and mechanic hiring information. Thanks to the helpful people at the AIA (Aerospace Industries Association of America) for the information on trends in the aerospace industry and also to Tom Horne and all the helpful people at the AOPA (Aircraft Owners and Pilots Association) for various addresses used herein. Special thanks to Rick Bogatko, Karen Lee, Danny Bruce, Peter Clark, Oscar McNeil, and G.D. Smith for their helpful insight in the career profiles. And last but not least, thanks to Diane, who spent countless hours typing and proofreading.

Introduction

At some time in our lives, we all find ourselves out in the job market trying to decide what we want to be "when we grow up." If planned properly, a professional career can be a rewarding experience. If not, it can make life seem long and uneventful.

The odds of finding a career that you will enjoy and excel at are much better when you take the time to prepare yourself for the hunt. Career planning does not have to be a painful experience. If tackled in an organized and planned manner, it can become an enjoyable (and of course rewarding) experience.

This book was designed to serve a dual purpose: first, it should provide the job hunter with a source of job information that can assist him in selecting a job field. Secondly, it provides a source of reference information complete with names and addresses that can be used for information and application.

Back when I graduated from college, I remember spending many hours researching various careers and potential employers. It doesn't take long to realize that you can't get hired if you're working in the library all the time. This book can save you much of that time so you can spend it on more worthwhile activities such as job interviewing.

1

Trends in the Aerospace Industry

Boeing T-43A

According to the Aerospace Industries Association of America, Inc. (AIA), employment in the aerospace industry was expected to reach 1,355,000 workers in 1987. Employment is down slightly from the 1986 statistics, but to put this into perspective, 1986 is surpassed only by the employment level attained in 1968 (the industry's peak employment year with 1,403,000 workers in the work force).

The aerospace industry has been expanding at a healthy rate, and the opportunities in this industry are among the best to be found.

As shown in Table 1-1, the estimated 1,355,000 employees in the 1987 workforce are broken down as follows: It is estimated that there were 585,000 production workers, 250,000 scientists and engineers, 104,000 technicians, and 416,000 various other employees in the industry.

One reason for the encouraging employment statistics is because the demand for aerospace products and its services is increasing. According to the AIA, estimated sales figures are increasing in terms of current dollars for civil aircraft, missiles, space research, and related products and services. Only military aircraft sales fail to remain ahead of the inflation rate. This trend will probably continue as one of the largest programs, the B-1B aircraft, draws to a close. See Table 1-2.

Table 1-1. AEROSPACE EMPLOYMENT (1968 THROUGH 1987)

		EMPLOYMENT BY OCCUPATIONAL TYPE (IN THOUSANDS)			
YEAR	TOTAL	PRODUCTION WORKERS	SCIENTISTS & ENGINEERS	TECHNICIANS	ALL OTHERS
1968	1,403	738	221	81	363
1969	1,295	658	203	72	362
1970	1,069	528	167	67	307
1971	924	448	159	60	257
1972	944	473	168	65	238
1973	962	484	164	66	248
1974	973	483	166	67	257
1975	925	444	167	63	251
1976	898	420	166	62	250
1977	894	410	173	59	252
1978	1,032	519	170	64	279
1979	1,152	592	177	69	314
1980	1,218	612	196	78	332
1981	1,203	578	194	84	347
1982	1,153	535	200	79	339
1983	1,171	528	207	87	349
1984	1,250	562	223[r]	93[r]	372[r]
1985	1,339	599	240	100	400
1986[(p)]	1,359	601	246	102	410
1987[(e)]	1,355	585	250	104	416

Source: Aerospace Industries Association, based on data from the U.S. Department of Labor, Bureau of Labor Statistics, "Employment and Earnings," and AIA Annual Survey of Aerospace Industry Employment.

[r] Revised
(p) Preliminary
(e) Estimate

The Department of Defense is by far the largest customer in the aerospace products and services industry. NASA and other customers also share in the total sales figures listed. See Table 1-3.

According to AIA statistics, defense-related aerospace sales are expected to grow at approximately 3 percent in real dollar terms in 1987 as the effect of existing programs balance the lower outlays expected in future budgets. Sales to NASA and other government agencies are expected to increase in the near future, but the longer-term outlook is for sales to level off because of the President's decision to take NASA and the federal government out of the commercial space business.

Table 1-2. Aerospace Industry Sales by Product Group, Calendar Years 1972 Through 1987 (Millions of Dollars)

Year	Total Sales	Aircraft Total	Aircraft Civil	Aircraft Military	Missiles	Space	Related
CURRENT DOLLARS							
1972	$23,610	$12,516	$ 4,181	$ 8,335	$ 4,285	$ 4,163	$ 2,646
1973	25,837	14,144	5,742	8,402	4,224	4,126	3,343
1974	27,454	14,867	6,320	8,547	4,108	4,412	4,067
1975	29,686	16,433	6,463	9,970	3,775	4,686	4,792
1976	29,825	16,056	6,007	10,049	3,671	4,787	5,311
1977	32,199	16,988	6,183	10,805	4,106	5,001	6,104
1978	37,702	21,074	8,222	12,852	4,098	5,717	6,813
1979	45,420	26,382	13,227	13,155	4,778	6,545	7,715
1980	54,697	31,464	16,285	15,179	6,469	7,945	8,819
1981	63,974	36,062	16,427	19,635	7,640	9,388	10,884
1982	67,756	35,484	10,982	24,502	10,368	10,514	11,390
1983	79,975	42,431	12,373	30,058	10,269	13,946	13,329
1984	83,486	41,905	10,690	31,215	11,335	16,332	13,914
1985	96,571	50,482	13,730	36,752	11,438	18,556	16,095
1986(p)	103,536	53,173	15,217	37,956	12,651	20,456	17,256
1987(e)	109,871	54,058	15,555	38,503	14,798	22,703	18,312
CONSTANT DOLLARS (Aerospace Composite Price Deflator, 1982 = 100)							
1972	$61,484	$32,594	$10,888	$21,706	$11,159	$10,841	$ 6,891
1973	60,226	32,970	13,385	19,585	9,846	9,618	7,792
1974	58,165	31,498	13,390	18,108	8,703	9,347	8,617
1975	56,011	31,005	12,194	18,811	7,123	8,842	9,041
1976	51,422	27,683	10,357	17,326	6,329	8,253	9,157
1977	51,850	27,356	9,957	17,399	6,612	8,053	9,829
1978	57,648	32,223	12,572	19,651	6,266	8,742	10,417
1979	62,822	36,490	18,295	18,195	6,609	9,052	10,671
1980	68,116	39,183	20,280	18,903	8,056	9,894	10,983
1981	70,768	39,891	18,171	21,720	8,451	10,385	12,040
1982	67,756	35,484	10,982	24,502	10,368	10,514	11,390
1983	76,239	40,449	11,795	28,654	9,789	13,295	12,706
1984	76,733	38,516	9,825	28,691	10,418	15,011	12,788
1985	87,872	45,934	12,493	33,441	10,408	16,885	14,645
1986(p)	91,141	46,807	13,395	33,412	11,136	18,008	15,190
1987(e)	93,269	45,890	13,205	32,685	12,562	19,272	15,545

Source: Aerospace Industries Association, based on data from the U.S. Department of Defense, National Aeronautics and Space Administration, U.S. Department of Commerce, Budget of the U.S. Government, company reports to AIA, and AIA estimates.

(p) Preliminary
(e) Estimate

Table 1-3. AEROSPACE INDUSTRY SALES BY CUSTOMER CALENDAR YEARS 1972 THROUGH 1987 (MILLIONS OF DOLLARS)

		Aerospace Products and Services				
Year	Total Sales	Total	Department of Defense	NASA and Other Agencies	Other Customers	Related Products and Services
CURRENT DOLLARS						
1972	$23,610	$20,964	$13,293	$ 5,022	$ 2,649	$ 2,646
1973	25,837	22,494	12,939	7,096	2,459	3,343
1974	27,454	23,387	12,638	8,141	2,608	4,067
1975	29,686	24,894	13,125	8,931	2,838	4,792
1976	29,825	24,514	13,403	8,173	2,938	5,311
1977	32,199	26,095	14,368	3,012	8,715	6,104
1978	37,702	30,889	15,533	3,151	12,205	6,813
1979	45,420	37,705	18,918	3,453	15,334	7,715
1980	54,697	45,878	22,795	4,106	18,977	8,819
1981	63,974	53,090	27,244	4,709	21,137	10,884
1982	67,756	56,366	34,016	4,899	17,451	11,390
1983	79,975	66,646	41,558	5,910	19,178	13,329
1984	83,486	69,572	45,969	6,063	17,540	13,914
1985	96,571	80,476	53,178	6,262	21,036	16,095
1986(p)	103,536	86,280	58,448	6,575	21,257	17,256
1987(e)	109,871	91,559	62,301	6,903	22,355	18,312
CONSTANT DOLLARS (Aerospace Composite Price Deflator, 1982 = 100)						
1972	$61,484	$54,593	$34,617	$ 6,898	$13,078	$ 6,891
1973	60,226	52,434	30,161	5,732	16,541	7,792
1974	58,165	49,548	26,775	5,525	17,248	8,617
1975	56,011	46,970	24,764	5,355	16,851	9,041
1976	51,422	42,265	23,109	5,065	14,091	9,157
1977	51,850	42,021	23,137	4,850	14,034	9,829
1978	57,648	47,231	23,751	4,818	18,662	10,417
1979	62,822	52,151	26,166	4,776	21,209	10,671
1980	68,116	57,133	28,387	5,113	23,633	10,983
1981	70,768	58,728	30,137	5,209	23,382	12,040
1982	67,756	56,366	34,016	4,899	17,451	11,390
1983	76,239	63,533	39,617	5,634	18,282	12,706
1984	76,733	63,945	42,251	5,573	16,121	12,788
1985	87,872	73,227	48,388	5,698	19,141	14,645
1986(p)	91,141	75,951	51,451	5,788	18,712	15,190
1987(e)	93,269	77,724	52,887	5,860	18,977	15,545

Source: Aerospace Industries Association, based on data from the U.S. Department of Defense, National Aeronautics and Space Administration, U.S. Department of Commerce, Budget of the U.S. Government, company reports to AIA, and AIA estimates.

(p) Preliminary

(e) Estimate

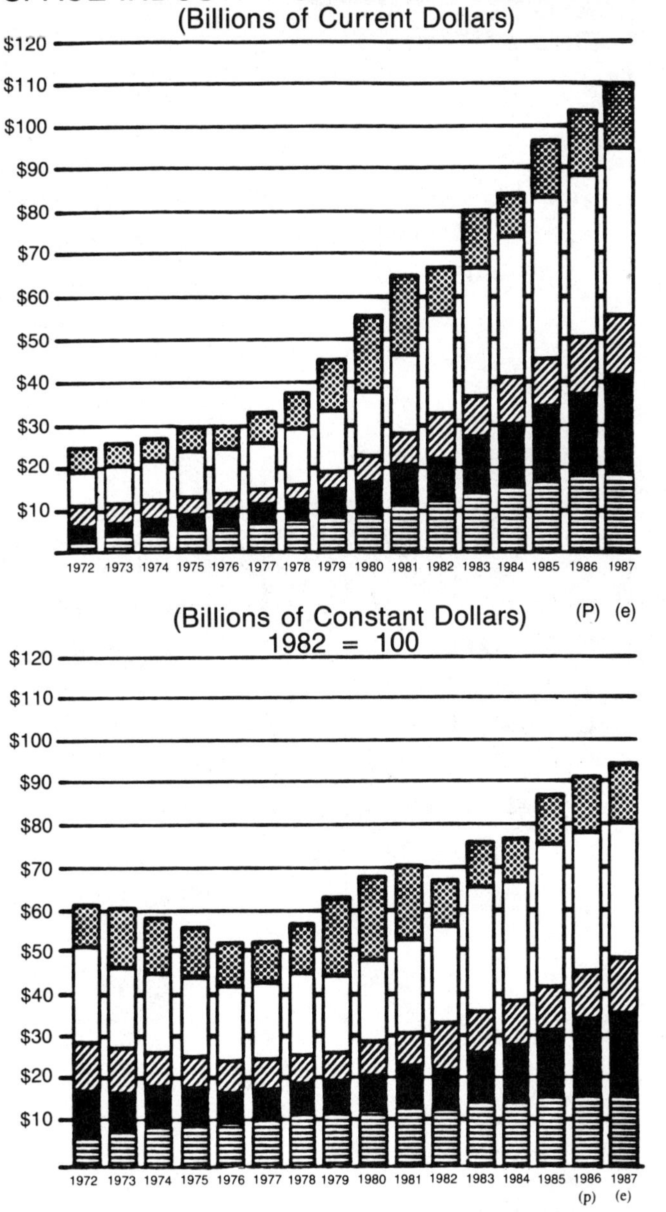

Reprinted with permission of the Aerospace Industries Association of America, Inc.

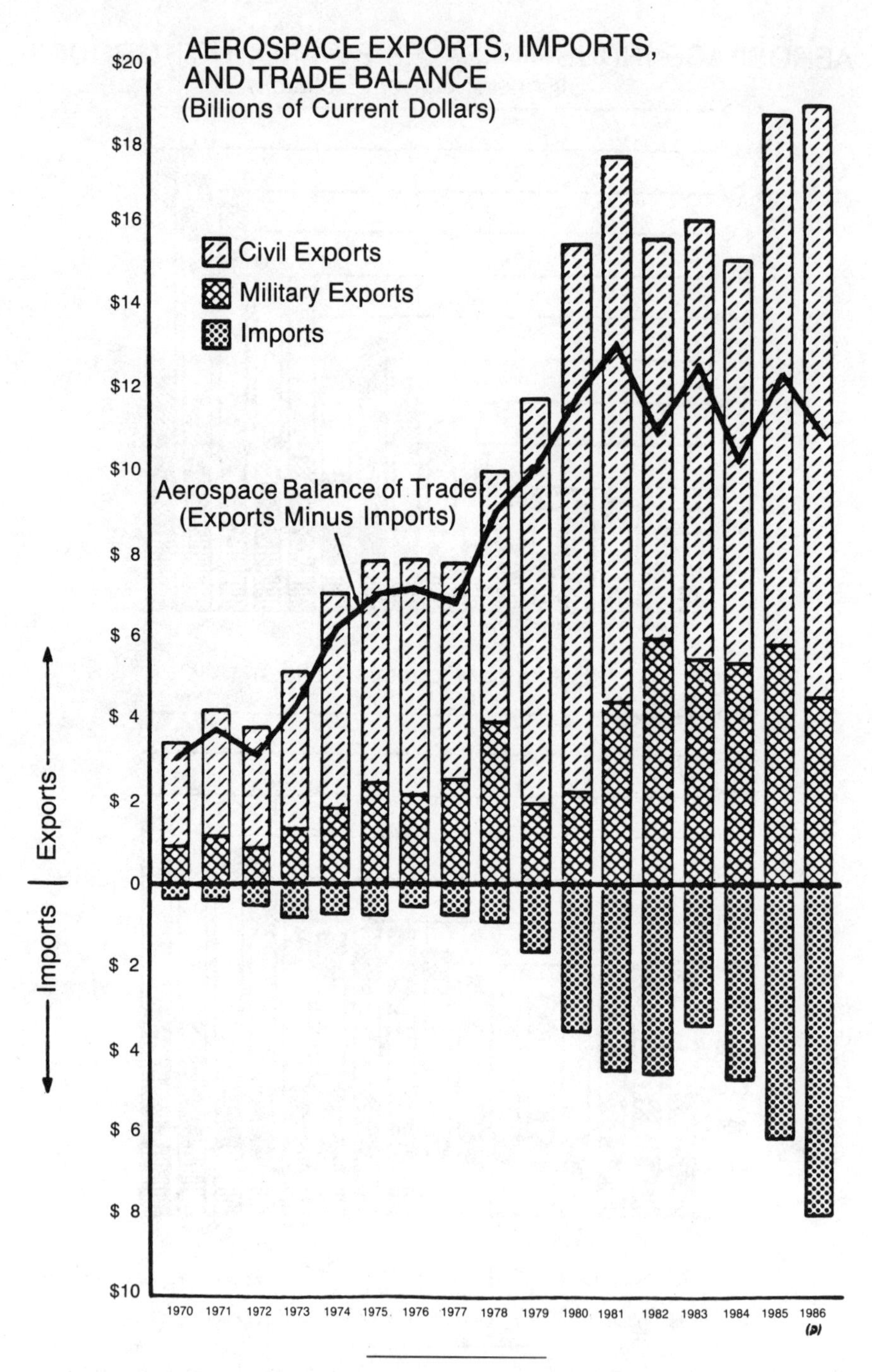

Reprinted with permission of the Aerospace Industries Association of America, Inc.

Civil aircraft sales continued to increase in 1986 but at a slower pace than in 1985. Commercial transport aircraft orders were responsible for the success of the commercial industry. The United States leads the world in the production of large commercial aircraft. Boeing and McDonnell Douglas now account for all U.S. production of jetliners since Lockheed left this market following delivery of its last L-1011 wide-body aircraft in 1985.

General aviation aircraft and helicopters show a substantial decrease in unit shipments. Because of rising product liability insurance rates and the abundance of serviceable, lower cost, used aircraft, new aircraft sales have dropped to the lowest level in years.

Due to this condition, manufacturers have reduced production of single-engine, piston-powered aircraft and concentrated on more profitable markets. Helicopter sales in 1986 were at the lowest level in many years. The depressed world oil market has had a great impact on helicopter production. Many oil companies use helicopters for transportation back and forth to oil rigs and other remote areas. Vertical take-off and landing (VTOL) capabilities make the helicopter essential for these offshore oil-drilling activities. Many oil companies purchased large numbers of helicopters when oil prices were high in the late 1970's and early 1980's. Because of this, many now find themselves with idle equipment. Until the oil market stabilizes, demand for this type transportation will probably continue to feel the impact.

In the near term, U.S. manufacturers will probably rely on the rescue, media, and police markets, because these areas hold the greatest potential for future sales. Military helicopter production will probably continue to boost the industry's overall performance. See Table 1-4.

According to the AIA, the United States by far exports more than it imports in the aerospace market. The United States exports aerospace products to both civil and military organizations in various foreign countries. Civil aircraft, engines, and spare parts account for approximately 76 percent of the total exports delivered. These civil exports include complete aircraft for transport, general aviation, helicopter and other markets, aircraft engines, and assorted aircraft repair parts and spares. Military exports account for approximately 24 percent of all exports and also include complete aircraft, aircraft engines, repair parts, and missiles. See Table 1-5.

Imports have risen sharply in the past few years (Table 1-6). Imports rose approximately 32 percent over 1985 figures. This is due largely to the increase in demand for small commuter aircraft. The demand for aircraft engines and repair parts also helped to boost the import trade balance figure. The increase in imports of aircraft engines is primarily due to the shipments of Rolls Royce engines for Harrier fighters, Gulfstream business jets, and Boeing 747 aircraft.

Aerospace employment is a function of supply and demand. The healthier the industry, the better the employment outlook. From all indications, the aerospace industry is alive and well in the United States and the outlook for employment in encouraging.

Table 1-4. Civil Aircraft Shipments, Calendar Years 1970 Through 1987

	NUMBER OF AIRCRAFT SHIPPED				VALUE (millions of dollars)			
Year	Total	Commerical Transport Aircraft[a]	Helicopters	General Aviation	Total	Commercial Transport Aircraft[a]	Helicopters	General Aviation
1970	8,076	311	482	7,283	$3,546	$ 3,158	$ 49	$ 339
1971	8,158	223	469	7,466	2,984	2,594	69	321
1972	10,576	227	575	9,774	3,308	2,660	90	558
1973	14,709	294	770	13,645	4,665	3,718	121	826
1974	15,326	332	828	14,166	5,091	3,993	189	909
1975	15,251	315	864	14,072	5,086	3,779	274	1,033
1976	16,429	222	757	15,450	4,592	3,078	285	1,229
1977	17,913	155	848	16,910	4,451	2,649	251	1,551
1978	18,962	241	904	17,817	6,458	4,308	328	1,822
1979	18,460	376	1,029	17,055	10,644	8,030	403	2,211
1980	13,634	387	1,366	11,881	13,058	9,895	656	2,507
1981	10,916	387	1,072	9,457	13,223	9,706	597	2,920
1982	5,085	232	587	4,266	8,610	6,246	365	1,999
1983	3,356	262	403	2,691[b]	9,773	8,000	303	1,470
1984	2,999	185	376	2,438	7,717	5,689	330	1,698
1985	2,683	278	376	2,029	10,436	8,500	505	1,431
1986(p)	2,030	329	301	1,400	11,807	10,534	273	1,000
1987(e)	2,106[c]	346	360	NA	12,241[c]	10,909	332	NA

Source: Aerospace Industries Association, based on company reports, data from the General Aviation Manufacturers Association, and AIA estimates.

[a] Includes all U.S. manufactured civil jet transport aircraft plus the turboprop-powered Lockheed L-100.
[b] Includes 3 off-the-shelf Gulfstream G-111's delivered to the U.S. Air Force for C-20 VIP transports.
[c] Due to the unavailability of general aviation forecast data, 1987 totals include 1986 general aviation figures for the purpose of estimating total shipments only.

NA Not available
(p) Preliminary
(e) Estimate

Table 1-5. Exports of U.S. Aerospace Products
Calendar Years 1981 Through 1986 (Millions of Dollars)

	1981	1982	1983	1984	1985	1986(p)
TOTAL EXPORTS	$17,634	15,603	$16,065	$15.008	$18,724	$18,988
TOTAL CIVIL	13,312	9,608	10,595	9,659	12,919	14,448
Complete Aircraft	8,613	4,848	5,691	4,147	6,694	7,085
Transports	7,180	3,834	4,683	3,195	5,518	6,000
General Aviation	790	517	356	268	191	251
Helicopters	346	206	232	234	210	260
Other, Including Used	297	291	420	450	775	574
Aircraft Engines	784	763	950	1,057	923	1,006
Turbine	739	721	914	1,021	880	965
Piston	45	42	36	36	43	41
Aircraft & Engine Parts Incl. Spares	3,915	3,997	3,954	4,455	5,302	6,357
TOTAL MILITARY	4,322	5,995	5,470	5,349	5,805	4,540
Complete Aircraft	1,712	2,388	1,845	1,581	2,011	1,176
Transports	158	341	112	85	101	2
Helicopters	177	156	62	83	117	142
Fighters & Bombers	1,006	1,473	1,379	977	1,352	800
Other, Including Used	371	418	292	436	441	232
Aircraft Engines	83	140	172	141	146	88
Turbine	78	136	162	125	144	85
Piston	5	4	10	16	2	3
Aircraft & Engine Parts Incl. Spares	1,971	2,341	2,459	2,665	2,823	2,611
Missiles, Rockets, & Parts	556	1,126	994	962	825	665

Source: Aerospace Industries Association, based on data from the U.S. Department of Commerce and AIA estimates.
Note: Import data include: non-military aircraft parts, and aerospace products previously exported from the U.S.

(p) Preliminary
r Revised

Table 1-6. IMPORTS OF U.S. AEROSPACE PRODUCTS CALENDAR YEARS 1981 THROUGH 1986 (MILLIONS OF DOLLARS)

	1981	1982	1983	1984	1985	1986(p)
TOTAL IMPORTS	$4,500	$4,568	$3,446	$4,926r	6,132	$8,102
Military Aircraft	42	28	3	14	20	37
Civil Aircraft	1,558r	1,599r	924r	1,301r	1,502	2,065
Transports	196	231	188	270	599	800
General Aviation	913	838	542	612	673	1,035
Helicopters	105	85	90	51	45	49
Other, Including Used Aircraft, Gliders, Balloons and Airships	344r	445r	104r	368r	185	181
Aircraft Engines	1,053	813	621	875	1,019	1,361
Turbine	1,048	803	605	862	1,011	1,351
Piston	5	10	16	13	8	10
Aircraft & Engine Parts Accessories & Equipment	1,847r	2,127r	1,898r	2,737r	3,374	4,639

Source: Aerospace Industries Association, based on data from the U.S. Department of Commerce and AIA estimates.
Note: Import data include: non-military aircraft parts, and aerospace products previously exported from the U.S.

(p) Preliminary
r Revised

2

Education and Training

Cessna C-172

As the aerospace industry becomes more dependent on high technology, so must the skills of the people who make the system work. Background education in many cases determines if a prospective employee is deemed qualified to perform the job in question.

There are many fine schools in this country that can prepare the applicant for an aviation profession. They are categorized as follows.

- ***Trade-Schools*** help to prepare the student to perform a specific kind of job. Many schools offer programs such as Airframe and Powerplant ("A & P") Mechanic training, avionics systems technician training, and flight training. These schools are relatively inexpensive.
- ***Junior Colleges*** typically offer an associate (two-year) degree. These programs allow the applicant to earn college credit that can be transferred (in most cases) to a fully accredited four-year college. Because most junior colleges are substantially less expensive than four-year colleges, this can sometimes equate to considerable savings for the student in the long run.
- ***Four-Year Colleges*** offer various programs of study ranging from avionics to airport management. Choosing the right school is one important step toward a successful career in aviation.

FAA-CERTIFICATED SCHOOLS

The FAA certified about 1,225 civilian flying schools, including some colleges and universities that offer degree credits for pilot training. Of the major airline pilots hired in 1988, 72.2 percent had four-year degrees while 9.9 percent had two to three years of college.

The remainder of this chapter is a listing of schools that are certificated by the Federal Aviation Administration. These schools are broken down into three major categories:

- ✦ schools for *pilots* (Table 2-1)
- ✦ schools for *aviation maintenance technicians* (Table 2-2)
- ✦ schools for *aircraft dispatchers* and *flight engineers* (Table 2-3)

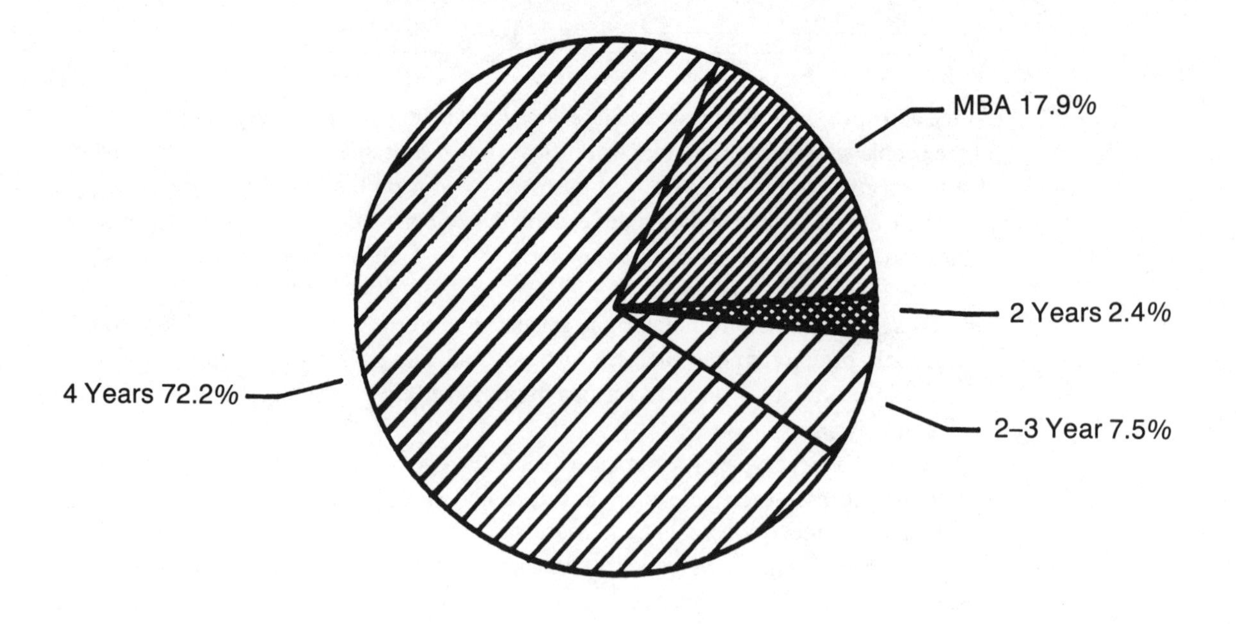

FAPA, Atlanta, GA 1-800-JET-JOBS

Certification of a school by the FAA indicates that the school meets the requirements of the Federal Aviation Regulations (FARs) for the ratings issued. These listings include the name of the school, its mailing address, and the ratings/courses offered. Schools are listed in alphabetical order within the state, U.S. possession, or territory in which they are located.

Additional copies of these listings can be obtained by writing to the following address:

SUPERINTENDENT OF DOCUMENTS
U.S. Government Printing Office
Washington, D.C. 20402-9325

For just the listing of pilot (Table 2-1) and flight engineer (Table 2-3) schools, request the *FAA Certificated Pilot Schools, Advisory Circular AC 140-2S* (cost at printing is $3.50). For the listing on aviation maintenance technician schools (Table 2-2) request the *FAA Certificated Aviation Maintenance Technician Schools, Advisory Circular AC 147-2Y* (cost at printing is $1.00).

In addition to these FAA listings, another good listing of college-level aviation/aerospace schools is *The Collegiate Aviation Directory*. This publication can be obtained by writing to the following address:

THE UNIVERSITY AVIATION ASSOCIATION
P.O. Box 2321
Auburn, AL 36830

(The cost at printing is $9.95.)

PILOT SCHOOLS

The following is a list of the codes used to designate ratings and courses offered by certificated pilot schools. The codes pertaining to each school are listed below its address.

Private Pilot Certification Course

PC(A)	Airplane
PC(R)	Rotorcraft
PC(G)	Glider
PC(L)	Lighter-Than-Air

Private Pilot Test Course

PT(A)	Airplane
PT(R)	Rotorcraft
PT(G)	Glider
PT(L)	Lighter-Than-Air

Commercial Pilot Certification Course

CC(A)	Airplane
CC(R)	Rotorcraft
CC(G)	Glider
CC(L)	Lighter-Than-Air

Commercial Pilot Test Course

CT(A)	Airplane
CT(R)	Rotorcraft
CT(G)	Glider
CT(L)	Lighter-Than-Air

Instrument Rating

IR(A)	Airplane
IR(H)	Helicopter

Test Preparation Courses

FI	Flight Instructor Certification
AF	Additional Flight Instructor Rating
AI	Additional Instrument Rating
AT	Airline Transport Pilot Certification
PR	Pilot Refresher Course
AG	Agricultural Aircraft Operations Course
XL	Rotorcraft External Load Operations Course

Additional Aircraft Rating

AR(A)	Airplane
AR(R)	Rotorcraft
AR(G)	Glider
AR(L)	Lighter-Than-Air

Pilot Ground School Courses

PG(P)	Private Pilot
PG(I)	Instrument Rating
PG(C)	Commercial Pilot
PG(F)	Flight Instructor
PG(T)	Airline Transport Pilot
PG(R)	Additional Flight Instructor
PG(A)	Additional Instrument

Table 2-1. FAA-CERTIFICATED PILOT SCHOOLS

ALABAMA

ALABAMA AVIATION AND TECH. COLLEGE
P.O. Box 1279
Hwy. 231 South
Ozark, AL 36360
PC(A),PT(A)
CC(A),CT(A),IR(A)
AR(A)(R)
AR(A)(R)
FI,AF,AI

AUBURN UNIVERSITY AVIATION
700 Airport Rd.
Auburn, AL 36830
PC(A),PT(A)
CC(A),IR(A)
AR(A)
FI,AI,AT

DIXIE AIR, INC.
P.O. Box 1370
Tuscaloosa, AL 35401
PC(A),PT(A)
CC(A),CT(A),IR(A)
AR(A)
FI,AF

EDWARDS AIRCRAFT INC.
Rt. 1, Box 26
Enterprise, AL 36330
PC(A)
CC(A),IR(A)
AR(A)
FI,AF,AI

GOLD DUST FLYING SERVICES INC.
305 Airport Rd.
Jacksonville Airport
Jacksonville, AL 36265
PC(A)
CC(A),IR(A)
AR(A),PG(P)(I)(C)

JACKSONVILLE STATE UNIVERSITY
Department of Physics
Jacksonville, AL 36265
PG(P)(I)(C)

NAPIER AIR SERVICE, INC.
Rt. 6, Box 206
Dothan, AL 36303
PC(A)
CC(A),IR(A)
AR(A)
FI,AF,AI,AT

REDSTONE ARSENAL FLYING CLUB
Building 118
Redstone Arsenal, AL 35809
PC(A)
CC(A),IR(A)
FI,AF

WALLACE STATE COMMUNITY COLLEGE
P.O. Box 250
Hanceville, AL 35077
PC(A)
CC(A),IR(A)
AR(A)
FI,AF,AI,AT

ALASKA

A I R CENTER
Tarbox, Wayne
P.O. Box 2528
Juneau, AK 99803
PC(A)
CC(A),IR(A)
AR(A),PG(P)(I)(C)

ACTION HELICOPTERS
507 W. Northern Lights Blvd.
Anchorage, AK 99503
PC(R)
CC(R),CT(R)
FI

AERO TECH FLIGHT SERVICE, INC.
1100 Merrill Field Dr.
Anchorage, AK 99501
PC(A),PT(A)
CC(A),CT(A),IR(A)
AR(A)
FI,AF,AT,PR

ALASKA AIR ACADEMY, INC.
2301 Merrill Field Dr.
Anchorage, AK 99501
PC(A)
CC(A),CT(A),IR(A)
AR(A)
FI,AF,AI,PR

ALASKA FLYING NETWORK
Rt. 2, Box 1621
Soldotna, AK 99669
PC(A)
CC(A),IR(A)

ALASKA WING
P.O. Box 60866
Fairbanks, AK 99706
PC(A)
IR(A)
PG(P)

ANCHORAGE AIR CENTER
1935 Merrill Field Dr.
Anchorage, AK 99501
PC(A)
CC(A),IR(A)
FI,AF

ANCHORAGE COMMUNITY COLLEGE
2533 Providence Ave.
Anchorage, AK 99504
PG(P)(I)(C)(F)(T)

AVIATION NORTH
Datco Realty, Inc.
P.O. Box 671528
Chugiak, AK 99567
PC(A)
CC(A)(R),CT(A),IR(A)
AR(R)
FI,AF

CLUB 1 FLIGHT GROUP
403 N. Willow
Kenai, AK 99611
PC(A)
CC(A),IR(A)
FI,AF

EIELSON AERO CLUB
Building 1121
Eielson AFB, AK 99702
PC(A),PT(A)
CC(A),CT(A),IR(A)
PG(P)(I)(C)(F)(R)
FI,AF

ELEMENDORF AERO CLUB
Hangar 7, Elemendorf AFB
Anchorage, AK 99506
PC(A)
CC(A)
AR(A),PG(P)(I)(C)(F)(T)(R)(A)
FI,AF,AT

FORT RICHARDSON FLYING CLUB
P.O. Box 5364
Anchorage, AK 99505
PC(A)
CC(A),IR(A)
AR(A)
FI,AF,AI,AT,PR

FORT WAINWRIGHT FLYING CLUB
P.O. Box 35062
Ft. Wainwright, AK 99703
PC(A),PT(A)
CC(A),CT(A),IR(A)
AR(A)
FI,AF

GORDON AVIATION, INC.
1931 Merrill Field Dr.
Anchorage, AK 99501
PC(A)
CC(A),CT(A),IR(A)
AR(A)
FI,AF,AT

HIGH TECH HELICOPTERS, INC.
Ballard, Wesley
2400 E. 5th Ave.
Anchorage, AK 99501
PC(R)
AR(R),PG(P)
FI

LARRY'S FLYING SERVICE
P.O. Box 2348
Fairbanks, AK 99707
PC(A)
CC(A),IR(A)
FI

LAZY MOUNTAIN AVIATION, INC.
P.O. Box 157
Palmer, AK 99645
PC(A)
CC(A),IR(A)
AR(A)
FI,AF

MAT SU BUSH FLYING, INC.
P.O. Box 2327
Palmer, AK 99645
PC(A)
CC(A),IR(A)
FI,AF

NORTH WIND AVIATION
Ballard, Wesley
2400 E. 5th Ave.
Anchorage, AK 99501
PC(A)
CC(A),IR(A)

PRECISION AVIATION FLYING CLUB
Brown, Carl L., Jr.
801 Airport Rd. Tra 1A
Palmer, AK 99645
PC(A),PT(A)

TANANA VALLEY COMMUNITY COLLEGE
Aviation Department
University of Alaska
P.O. Box 7
Fairbanks, AK 99701
PG(P)

THE AVIATION CO.
Snyder, Ken
3762 S. University Ave.
Fairbanks, AK 99709
PC(A),PT(A)
CC(A),CT(A)(R),IR(A)
AR(A)(R),PG(P)
FI,AT

VERNAIR
1704 E. 5th Ave.
Anchorage, AK 99501
PC(A),PT(A)
CC(A),IR(A)
AR(A)
FI,AF

WILBURS FLIGHT OPERATIONS
1740 E. Fifth Ave.
Anchorage, AK 99501
PC(A)
CC(A),IR(A)(H)
AR(A)(R)
FI,AF,AT

ARIZONA

CHANDLER AIR SERVICE, INC.
1675 E. Ryan Rd.
Chandler, AZ 85249
PC(A)
CC(A),IR(A)

COCHISE COMMUNITY COLLEGE
Rt. 1, Box 100
Douglas, AZ 85607
PC(A)
CC(A)
AR(A)
FI,AF

EMBRY-RIDDLE AERONAUTICAL UNIVERSITY
3200 N. Willow Creek Rd.
Prescott, AZ 86301
CC(A)

FLIGHTSAFETY INTERNATIONAL, INC.
6870 S. Plumer Ave.
Tucson, AZ 85706
AR(A)
AT,PR

HUDGIN AIR SERVICE
1732 E. Valencia
Tucson, AZ 85706
PC(A)
CC(A),IR(A)
AR(A)
FI,AF,AT

LITCHFIELD AVIATION
Phoenix-Litchfield Airport
Goodyear, AZ 85338
PC(A)
CC(A),IR(A)
AR(A)
FI

NORTH AIRE, INC.
Prescott Municipal Airport
Prescott, AZ 86301
PC(A)
CC(A),IR(A)
FI,AF

SAWYER SCHOOL OF AVIATION
Sawyer Aviation Inc.
2602 E. Sky Harbor Blvd.
Phoenix, AZ 85034
PC(A),PT(A)
CC(A),CT(A),IR(A)
AR(A)
FI,AF,AT

SUPERSTITION AIR SERVICE
4766 Falcon Dr.
Mesa, AZ 85205
PC(A)
CC(A),IR(A)
AR(A)
FI,AF

ARKANSAS

CENTRAL FLYING SERVICE, INC.
Adams Field
Little Rock, AR 72202
PC(A),PT(A)
CC(A),CT(A),IR(A)
AR(A),PG(P)(I)(C)
FI,AF,AT

MOREY'S FLYING SERVICE
Morey, Danny J
Rt. 2, Box 48
Rogers, AR 72756
PC(A)
CC(A),IR(A)
AR(A)
FI,AF

MUSTANG, INC.
3901 Lindberg
Jonesboro, AR 72401
PC(A)
CC(A)

CALIFORNIA

ABOVE ALL AVIATION
1501 Cooke Pl
Goleta, CA 93117
PC(A)
CC(A),IR(A)

ACCELERATED GROUND TRAINING, INC.
19531 Airport Way South
Santa Ana, CA 92707
PG(P)(I)(C)(F)(T)(R)(A)

AERO TECH ACADEMY
Hovell, John L.
1745 Sessums Dr.
Redlands, CA 92373
PC(A)
CC(A),IR(A)
FI,AF

AFTA INC.
4773 Highland Springs Rd.
Lakeport, CA 95453
PC(A),PT(A)
CC(A),CT(A),IR(A)

AIR TRAILS, INC.
280 Mortensen Ave.
Salinas, CA 93905
PC(A)
CC(A),CT(A),IR(A)
AR(A)
FI,AF

AIRFLITE, INC.
2700 E. Wardlow Rd.
Long Beach, CA 90807
PC(A),PT(A)
CC(A),CT(A),IR(A)
AR(A)
FI,AF

AIRLINE TRAINING INSTITUTE
Skyway Aviation Inc.
795 Skyway
San Carlos, CA 94070
PC(A)
CC(A),IR(A)
FI,AF,AT

AIRSOUTH
4119 W. Commonwealth Ave.
Fullerton, CA 92663
PC(A),PT(A)
CC(A),CT(A),IR(A)
AR(A)
FI,AF

ALL AMERICAN FLIGHT CENTER
Davis, Cornelia M
1935 N. Marshall
El Cajon, CA 92020
PC(A)
CC(A),IR(A)
FI,AF

AMERICAN AEROBATICS
820-D E. Santa Maria St.
Santa Paula, CA 93060
PR

AMERICAN EAGLE AVIATION
20915 Lamberton Ave.
Long Beach, CA 90810
PG(P)(I)(C)(F)(R)(A)
FI,AF,AI

AMERICAN FLYERS
ATE of California Inc.
3021 Airport Ave.
Santa Monica, CA 90405
PC(A)
CC(A),IR(A)
AR(A)
FI,AF,AT,PR

ARIS HELICOPTERS LTD.
1138 Coleman Ave.
San Jose, CA 95110
PC(R)
CC(R),IR(H)
AR(R)
FI,AF,AI,AT,XL

AVIATION TRAINING, INC.
21593 Skywest Dr.
Hayward, CA 94541
PC(A)
CC(A),IR(A)

BALLOON EXCELSIOR SCHOOL
Stockwell, Brent
1241 High St.
Oakland, CA 94601
AR(L)

BALLOONS OF WOODLAND
Wallis, John A.
1233 E. Beamer
Suite E1
Woodland, CA 95695
CC(L)

BATES FOUNDATION FOR AERONAUTICAL EDUCATION
Harvey Mudd College
Claremont, CA 91711
PC(A)
CC(A),IR(A)

BERG BRANHAM FLYING SERVICE, INC.
16425 Hart St.
Van Nuys, CA 91406
PC(A)
CC(A),IR(A)
AR(A)
FI,AF,AT

BRIDGEFORD FLYING SERVICE, INC.
R. A. Bridgeford, Inc.
Napa County Airport
Napa, CA 94558
PC(A)
CC(A),IR(A)
AR(A),PG(P)(I)
FI,AF,AT

CAL-AG-AERO
Calif. Agricultural Aeronautics, Inc.
P.O. Box 939
Tulare, CA 93275
AG

CALIFORNIA AVIATION
C and K Aviation Inc.
2501 Airport Ave.
Santa Monica, CA 90405
PC(A),PT(A)
CC(A),IR(A)
AR(A)

CALIFORNIA AVIATION, INC.
2701 Airport Ave.
Santa Monica, CA 90405
PC(A)
CC(A),CT(A),IR(A)
AR(A),PG(P)
FI,AF,AT

CALIFORNIA FLIGHT SCHOOL
A V Aviation, Inc.
William J. Fox Field
4555 W. Ave. G
Lancaster, CA 93534
PC(A)(R)
AR(R)

CALIFORNIA WINGS, INC.
4025 Kearny Villa Rd.
San Diego, CA 92123
PC(A)
CC(A),IR(A)
AR(A)
FI,AF

CAPITOL SKY PARK FLYING SCHOOL
Capitol Sky Park Inc.
Executive Airport
Sacramento, CA 95822
PC(A),PT(A)
CC(A),CT(A),IR(A)
AR(A)
FI,AF,AT

CHANNEL ISLANDS AVIATION, INC.
Camarillo Airport
Durley Ave., Building 233
Camarillo, CA 93010
PC(A)
CC(A),IR(A)
AR(A)
FI,AF

COLLEGE OF SAN MATEO
1700 W. Hillsdale Blvd.
San Mateo, CA 94402
PG(P)(I)(C)

COLLEGE OF THE REDWOODS
Richter, L.E.
Tompkins Hill Rd.
Eureka, CA 95501
PG(P)(I)(C)

Consolidated Aviation Corp.
Compton Aviation Corp.
Pal Aviation Intl. Inc.
961 W. Alondra Blvd.
Compton, CA 90220
PC(A)
CC(A),IR(A)

Eagle Airlines
Eagle Aviation Inc.
4307 Donald Douglas Dr.
Long Beach, CA 90808
PC(A)
CC(A),IR(A)
AR(A)
FI,AF,AT

EL Cajon Flying Service, Inc.
1825 N. Marshall Ave.
El Cajon, CA 92020
PC(A)
CC(A),IR(A)
AR(A)
FI,AF,AT

EL Monte Flight Service
5001 N. Santa Anita Ave.
El Monte, CA 91731
PC(A)
CC(A),IR(A)
AR(A)(R)
FI,AF

El Toro Marine Aero Club
MCAS El Toro
Santa Ana, CA 92709
PC(A)
CC(A),IR(A)
AR(A)
FI,AF

Executive Aero Systems, Inc.
P.O. Box 11557
Tahoe Paradise, CA 95708
PC(A),PT(A)
CC(A),CT(A),IR(A)

Executive Aviation Services
Thorson Enterprises
3521 E. Spring St.
Long Beach, CA 90806
AR(A)

Fallbrook Air Service
Aberle, Yvonne J
2141 S. Mission Rd.
Fallbrook, CA 92028
PC(A)

Flight Associated Activities
Flighteast, Inc.
6920 Vineland Ave.
North Hollywood, CA 91605
PC(A)

Flight Trails A California Corp.
2188 Palomar Airport Rd.
Carlsbad, CA 92008
PC(A)(R)
CC(A)(R),IR(A)(H)
AR(A)(R)
FI,AF,AT

Flying Country Club Marketing, Inc.
2555 Robert Fowler Way A
San Jose, CA 95148
PC(A),PT(A)
CC(A),CT(A),IR(A)
AR(A)
FI,AF

Flying J Aviation
Jones, Billy R
6717 Curran St.
San Diego, CA 92173
PC(A)
CC(A),IR(A)
AR(A),PG(P)(I)(C)(F)(R)
FI,AF

Foothill College
12345 El Monte Rd.
Los Altos Hills, CA 94022
PG(P)(I)

Fullerton Skyways, Inc.
3815 W. Commonwealth Ave.
Fullerton, CA 92633
PC(A),PT(A)
CC(A),CT(A)
FI

GENERAL AIR SERVICES, INC.
Graham, J. M.
260 Buchanan Field Rd.
Concord, CA 94520
PC(A)
CC(A),IR(A)
AR(A)
FI,AF

GIBBS FLITE CENTER, INC.
3717 John J. Montgomery Dr.
San Diego, CA 92123
PC(A)
CC(A),IR(A)
AR(A)(R),PG(P)(I)(C)(T)
FI,AF,AT,XL

GLENDALE COMMUNITY COLLEGE
1500 N. Verdugo Rd.
Glendale, CA 91208
PG(P)

GOLDEN GATE PIPER
Sudon Corp.
603 Skyway
San Carlos, CA 94070
PC(A),PT(A)
CC(A),CT(A),IR(A)
AR(A)
FI,AF,AT

GOLDEN STATE AVIATION
Golden State Flying Club
Volar Inc.
1640 N. Johnson Ave.
El Cajon, CA 92020
PC(A)
PG(P)

GROSSMOUNT COMMUNITY COLLEGE
8800 Grossmount College Dr.
El Cajon, CA 92020
PG(P)(I)(C)

GUNNELL AVIATION, INC.
3000 Airport Ave.
Santa Monica, CA 90405
PC(A),PT(A)
CC(A),CT(A),IR(A)
AR(A),PG(R)
FI,AT

HARBOR AVIATION
Harbor Enterprises, Inc.
4225 Donald Douglas Dr.
Long Beach, CA 90808
PC(A)
CC(A),IR(A)
AR(A)
FI,AF

HELIFLIGHT SYSTEMS, INC.
3205 Lakewood Blvd.
Long Beach, CA 90808
PC(R)
CC(R),IR(H)
PG(P)(C)
FI

IASCO FLIGHT TRAINING CENTER
International Air Service Co., Ltd.
100 Iasco Rd.
Napa County Airport
Napa, CA 94558
PC(A),PT(A)
CC(A),CT(A),IR(A)
AR(A),PG(P)(I)(C)
FI,AF,AI,AT,PR

KARLS FLYING SERVICE
Harder, C. Karl
P.O. Box 607
Lincoln Airport
Lincoln, CA 95648
PC(A)
CC(A),IR(A)
AR(A)
FI,AF,AT

KINGSBURY AVIATION
General Aviation Co.
3915 W. Commonwealth Ave.
Fullerton, CA 92633
PC(A),PT(A)
CC(A),CT(A),IR(A)
AR(A)
FI,AF,AI,AT

LAKE TAHOE AVIATION, INC.
P.O. Box 7323
South Lake Tahoe, CA 95731
PC(A),PT(A)

LANCER AVIATION, INC.
1900 Joe Crosson Dr.
El Cajon, CA 92020
PC(A)
PG(P)

LENAIR AVIATION, INC.
Lackey, Gary Wayne
19531 Airport Way South
Suite 5
Santa Ana, CA 92707
PC(A)
CC(A),IR(A)
AR(A)
FI,AT

LONG BEACH FLYERS
2901 E. Spring St.
Long Beach, CA 90806
PC(A)
CC(A),IR(A)
AR(A)
FI,AF,AI

MARIN AIR SERVICES
P.O. Box 1828
Novato, CA 94948
PC(A)
CC(A),IR(A)
AR(A)
FI,AF,AT

MAZZEI FLYING SERVICE
Golden Eagle Enterprises, Inc.
4955 E. Anderson
Fresno, CA 93727
PC(A)(R),PT(A)
CC(A),CT(A),IR(A)
AR(A)(R)
FI,AF,AT

NATIONAL AIR COLLEGE, INC.
3760 Glenn H. Curtis Rd.
San Diego, CA 92071
PC(A),PT(A)
CC(A)(R),CT(A)(R),IR(A)
AR(A)(R),PG(P)(I)(C)(F)(T)(R)
FI,AF,AT

NATIONAL UNIVERSITY
4141 Camino Del Rio South
San Diego, CA 92108
PC(A)
CC(A),IR(A)
AR(A),PG(P)(I)(C)(F)(T)(R)
FI,AF,AT

NAVAJO AVIATION
Laron Enterprises Inc.
145 John Glenn Dr.
Concord, CA 94520
PC(A),PT(A)
CC(A),CT(A),IR(A)
AR(A)
FI,AF,AT

NORTH ISLAND NAVY FLYING CLUB
P.O. Box 12
Nas North Island
San Diego, CA 92135
PC(A)
CC(A),IR(A)
AR(A)
FI,AF

OCEANAIR FLIGHT SERVICES, INC.
480 Airport Rd.
Oceanside, CA 92054
PC(A)
CC(A),IR(A)

PACIFIC AIR COLLEGE
Long Beach Flyers, Inc.
2901 E. Spring St.
Long Beach, CA 90808
PG(P)(I)(C)

PACIFIC FLIGHT AVIATION
6775 Airport Dr.
Riverside, CA 92504
PC(A)
CC(A),IR(A)

PACIFIC STATES AVIATION, INC.
Boggess, William
51 John Glenn Dr.
Concord, CA 94524
PC(A)
CC(A),IR(A)
FI,AF

PARFLITE, INC.
6651 Flight Rd.
Riverside, CA 92504
PC(A)

PASADENA CITY COLLEGE
1570 E. Colorado Blvd.
Pasadena, CA 91106
PG(P)(I)(C)

PATTERSON AIRCRAFT
Executive Airport
Sacramento, CA 95822
PC(A)
CC(A),IR(A)
AR(A)
FI,AF,AT

PHILLIPS FLYING
109 Crystal Air
South Lake Tahoe, CA 95705
PG(P)

PROFESSIONAL AIRLINE SYSTEMS
Pappas, Michael C.
2141 Palomar Airport Rd.
Carlsbad, CA 92008
AR(A)

RAMAIR CORP.
2735 E. Spring St.
Long Beach, CA 90806
PC(A)
CC(A),IR(A)
AR(A)
FI,AF,AI,AT

ROSE AVIATION, INC.
3852 W. 120th St.
Hawthorne, CA 90250
PC(A)
CC(A),IR(A)
PG(P)(I)(C)(F)(T)(R)

ROSEMEAD ADULT SCHOOL
9063 E. Mission Dr.
Rosemead, CA 91770
PG(P)

RYAN AIRE, INC.
620 Airport Dr.
Suite 5
San Carlos, CA 94070
PC(A),PT(A)
CC(A),CT(A),IR(A)

SAN BERNARDINO VALLEY COLLEGE
701 S. Mt. Vernon Ave.
San Bernardino, CA 92410
PG(P)

SAN CARLOS FLIGHT CENTER
620 Airport Dr.
San Carlos, CA 94070
PC(A)
CC(A),IR(A)

SAN DIEGO MESA COLLEGE
7250 Mesa College Dr.
San Diego, CA 92111
PC(A)
CC(A),IR(A)

SANTA ANA COLLEGE
17th At Bristol St.
Santa Ana, CA 92706
PG(P)(I)(C)

SANTA ROSA JUNIOR COLLEGE
Ryan, Dennis M.
1501 Mendocino Ave.
Santa Rosa, CA 95401
(PG)(P)

SIERRA ACADEMY OF AERONAUTICS
Everett, N. N.
130 Earhart Rd., Building L
Oakland Intl. Airport
Oakland, CA 94614
PC(A)
CC(A),IR(A)
AR(A)(R)
FI,AF,AT,XL

SKYHAWK AVIATION, INC.
Corona Municipal Airport
Corona, CA 91720
PC(A),PT(A)

SOUTHWEST SKYWAYS, INC.
25321 Bellanca Way
Torrance, CA 90505
PC(A)
CC(A),IR(A)
AR(A)
FI,AF

STAR AIRCRAFT SALES, INC.
1900 Joe Crosson Dr.
Gillespie Field
El Cajon, CA 92020
PC(A)

STECK AVIATION, INC.
6949 Curran St.
Brown Field
San Diego, CA 92173
PC(A),PT(A)
CC(A),IR(A)
AR(A),PG(P)
FI,AF

SWEETWATER COMMUNITY COLLEGE
900 Otay Lakes Rd.
Chula Vista, CA 92071
PG(P)(I)(C)

TRANS BAY AIRWAYS CORP.
620 Airport Dr.
Suite 1
San Carlos, CA 94070
PC(A),PT(A)
CC(A),CT(A),IR(A)

TRAVIS AFB AERO CLUB
West Hanger Ave.
P.O. Box 1477
Travis AFB, CA 94535
PC(A)
CC(A),IR(A)
FI,AF,AI,AT

VALENTI AVIATION
Valenti Aviation Inc.
1601 W. Fifth St.
Oxnard, CA 93030
PC(A)
CC(A),CT(A),IR(A)
AR(A)
FI,AF,AT

VIAIR, INC.
200 N. El Cielo
Palm Springs, CA 92262
PC(A)
CC(A),IR(A)

VINDAR AVIATION, INC.
Scott, Darrell J.
P.O. Box 747
Novato, CA 94947
PC(A)
CC(A),IR(A)
AR(A)
FI,AF,AT

WESTERN HELICOPTERS, INC.
1670 Miro Way
P.O. Box 579
Rialto, CA 92376
PC(R),PT(R)
CC(R),CT(R)

WESTERN SUN AVIATION, INC.
2025 N. Marshall Ave.
El Cajon, CA 92020
PC(A)
CC(A),IR(A)
AR(A)
FI,AF

COLORADO

AIMS COMMUNITY COLLEGE
5401 W. 20th St.
Greeley, CO 80632
PG(P)(I)(C)

COLORADO NORTHWESTERN COMMUNITY COLLEGE
Rangely Airport
Rangely, CO 81648
PC(A)
CC(A),IR(A)
FI

DENVER AIR CENTER, INC.
Jefferson County Airport
Building B3
Broomfield, CO 80020
PC(A)
CC(A),IR(A)

DURANGO AIR SERVICE, INC.
P.O. Box 2117
La Plata County Airport
Durango, CO 81301
PC(A),PT(A)
CC(A),CT(A),IR(A)
AR(A)

EMERY SCHOOL OF AVIATION
Johnson and Johnson Education Sys.
661 Buss Ave.
Greeley, CO 80631
PC(A)
CC(A)(R),IR(A)
AR(A)(R)
FI,AF,AT

FLOWER AVIATION OF COLORADO, INC.
31201 Bryan Circle
Pueblo, CO 81001
PC(A),PT(A)

HOFFMAN PILOT CENTER, INC.
Executive Office Building
Suite 7
Jeffco Airport Industrial Park
Broomfield, CO 80020
PC(A)(R)
CC(A)(R),IR(A)(H)
AR(R)
FI,AF,AI,AT,PR,XL

JUDSON FLYING SCHOOL
Rt. 3, Box 121
Longmont, CO 80501
PC(A)
CC(A),IR(A)
AR(A)
FI,AF

METROPOLITAN STATE COLLEGE
1006 11th St.
Denver, CO 80204
PG(P)(I)(C)

MONARCH AVIATION, INC.
Walker Field
Grand Junction, CO 81501
PC(A)
CC(A),CT(A),IR(A)
AR(A)
FI,AF,AI,PR

NORTHERN COLORADO AVIATION
780 Buss Ave.
Greeley, CO 80631
PC(A)
CC(A),IR(A)
AR(A),PG(F)(R)

P C FLYERS
7355 Peoria St.
Englewood, CO 80112
PC(A)
CC(A),IR(A)

PETERSON AFB AERO CLUB
P.O. Box 14123, Building 104
Peterson AFB, CO 80914
PC(A)
CC(A),IR(A)
AR(A)
FI,AF,AT

PROPILOT PROFICIENCY CENTER, INC.
715 Locust St.
Denver, CO 80220
PR

ROCKY MOUNTAIN PIPER
Jefferson County Airport
Building 4
Broomfield, CO 80020
PC(A),PT(A)
CC(A),CT(A),IR(A)
AR(A)
FI,AF

UNITED AIRLINES FLIGHT TRAINING CENTER
United Airlines
Stapleton Intl. Airport
Denver, CO 80207
AR(A),PG(T)

WEST AIRE, INC.
1245 Aviation Way
Colorado Springs, CO 80916
PC(A),PT(A)
CC(A),CT(A),IR(A)
FI

WINGS OF DENVER FLYING CLUB, INC.
7625 S. Peoria St.
Box 10
Englewood, CO 80112
PC(A)
CC(A),IR(A)
FI,AF,AT

CONNECTICUT

BLUEBIRD AVIATION CORP.
Danbury Municipal Airport
Danbury, CT 06810
PC(A)
CC(A),IR(A)

CHESTER AIRPORT, INC.
Winthrop Rd.
Chester, CT 06412
PC(A),PT(A)
CC(A),CT(A),IR(A)
FI,AF

COASTAL AIR SERVICES, INC.
Groton-New London Airport
Groton, CT 06340
PC(A)
CC(A),IR(A)
AR(A)
FI,AF,AT

INTERSTATE AVIATION, INC.
82 Johnson Ave.
Plainville, CT 06062
PC(A)
CC(A),IR(A)
FI,AF

KELAIRE, INC.
Sikorsky Memorial Airport
Stratford, CT 06497
PC(A)

MERIDEN AIRWAYS
Air Service, Inc.
213 Evansville Ave.
Meriden, CT 06450
PC(A)
CC(A),IR(A)
FI,AF

NORTHEAST FLIGHT TRAINING CENTER
Northeast Helicopters, Inc.
Hangar Two
Ellington Airport
Ellington, CT 06029
PC(A)(R)
CC(A)(R),IR(A)
AR(R)

SHORELINE AVIATION, INC.
1362 Boston Post Rd.
Madison, CT 06443
PC(A),PT(A)
CC(A),CT(A),IR(A)

STAPLES HIGH SCHOOL
70 North Ave.
Westport, CT 06880
PG(P)

WATERFORD FLIGHT SCHOOL
New London Flying Service, Inc.
Waterford Airport
Waterford, CT 06385
PC(A),PT(A)
CC(A),CT(A),IR(A)
AR(A)
FI,AF,AT

WINDHAM AVIATION, INC.
P.O. Box 136
Willimantic, CT 06226
PC(A),PT(A)
CC(A),CT(A),IR(A)
AR(A)
FI,AF

DELAWARE

DAWN AERONAUTICS, INC., NR 2
120 Old Churchmans Rd.
Greater Wilmington Airport
New Castle, DE 19720
PC(A)
CC(A),IR(A)
AR(A)
FI,AF,AT

DELAWARE TECHNICAL AND COMMUNITY COLLEGE
Barrett, Lowell A.
Terry Campus
1832 N. Dupont Pkwy.
Dover, DE 19901
PG(P)(I)(C)

FLIGHTSAFETY INTERNATIONAL, INC.
P.O. Box 15003
Greater Wilmington Airport
Wilmington, DE 19850
AR(A)

KENT COUNTY VO. TECH. CENTER
P.O. Box 97
Woodside, DE 19980
PG(P)(I)(C)

FLORIDA

ACADEMY OF FLIGHT
Orlando Aviation, Inc.
4800 E. Colonial Dr.
Orlando, FL 32803
PC(A),PT(A)
CC(A),CT(A),IR(A)
PG(P)(I)(C)

AERO SPORT FLIGHT CENTER, INC.
P.O. Box 1719
St. Augustine Airport
St. Augustine, FL 32084
PC(A)
CC(A),IR(A)
AR(A)(R),PG(P)(I)(C)(F)(R)
FI,AF,AT,PR

AEROSERVICE INTERNATIONAL TRAINING CENTER, INC.
1499 NW 79th Ave.
Miami, FL 33126
AR(A)
AT

AIR SANLANDO, INC.
Building 332
Sanford Central Florida Airport
Sanford, FL 32771
PC(A)
CC(A),IR(A)
AR(A)
FI,AF

AIR VENICE, INC.
P.O. Box 637
Venice, FL 33595
PC(A)

AMERICAN FLYERS
ATE of Florida
5500 NW 21, Terrace Building 4
Ft. Lauderdale, FL 33309
PC(A)
CC(A),IR(A)
PG(P)(I)(C)

ATE OF FLORIDA, INC.
1006 NE 11th St.
Pompano Beach, FL 33060
PC(A)
CC(A),IR(A)
FI,AF

BARTOW FLYING SERVICE
Bartow Municipal Airport
Development Auth.
P.O. Box 650
Bartow, FL 33830
PC(A)
IR(A)
PG(P)
FI

BAY AVIATION
Bay Aero Center, Inc.
1000 Jackson Way
Panama City, FL 32405
PC(A),PT(A)
CC(A),CT(A),IR(A)
AR(A)

BBR AVIATION, INC.
2405 SE Dixie Hwy.
Stuart, FL 33494
PC(A)

BOCA FLIGHT CENTER, INC.
3900 Perimeter Rd.
Boca Raton, FL 33431
PC(A),PT(A)
CC(A),CT(A),IR(A)
PG(P)(I)(C)

BRAUNIG AEROMARINE
1832 Spruce Creek Blvd. E.
Daytona Beach, FL 32014
PC(A)
CC(A),IR(A)
AR(A),PG(P)(I)(C)(T)
AT

BROWARD COMMUNITY COLLEGE
7200 Hollywood Blvd.
Pembroke Pines, FL 33024
PC(A)
CC(A),IR(A)
PG(P)(I)(C)

CARDINAL AVIATION, INC.
Moorman, Paul A.
301 Lexington Ave.
Deland Municipal Airport
Deland, FL 32720
PC(A)

CARIB AVIATION, INC.
Pules, Paul
14250 SW 129th St.
Miami, FL 33186
PC(A)
CC(A),IR(A)
PG(P)(I)(C)

CAV AIR, INC.
5500 NW 21st Terrace
Hangar 9
Ft. Lauderdale, FL 33309
PC(A)
CC(A),IR(A)
AR(A),PG(P)(I)(C)(F)

CORPORATE AIRWAYS
P.O. Box 18341
Jacksonville Intl. Airport
Jacksonville, FL 32229
PC(R)
CC(R)
AR(R)
FI

CRESCENT AIRWAYS, INC.
7501 Pembroke Rd.
W. Hollywood, FL 33023
AR(R)
FI,AF

CRYSTAL AERO GROUP, INC.
Davis, Thomas E.
P.O. Box 2050
Crystal River, FL 32629
PC(A),PT(A)
CC(A),CT(A),IR(A)

DAYTONA BEACH AVIATION, INC.
561 Pearl Harbor Dr.
Regional Airport
Daytona Beach, FL 32014
PC(R)
CC(R),IR(H)
AR(R),PG(P)(I)(C)(F)

EGLIN AERO CLUB
P.O. Box 1588
Hangar 66
Eglin AFB, FL 32542
PC(A)
CC(A),IR(A)
AR(A)
FI,AF,AT

EMBRY-RIDDLE AERONAUTICAL UNIVERSITY
Regional Airport
Daytona Beach, FL 32014
PC(A)
CC(A),IR(A)
AR(A),PG(P)(I)(C)(F)(R)
FI,AF

FERGUSON FLYING SERVICE, INC.
Ferguson Airport
Rt. 2, Box 935
Pensacola, FL 32506
PC(A)
CC(A),IR(A)
AR(A)(R)
FI,AF

FLIGHTSAFETY INTERNATIONAL, INC.
P.O. Box 2708
Vero Beach, FL 32960
PC(A)(R)
CC(A)(R),IR(A)(H)
AR(A)(R)
FI,AF,AI,AT,PR

FLORIDA INSTITUTE OF TECHNOLOGY
640 Harry Sutton Rd.
Melbourne, FL 32901
PC(A)
CC(A),IR(A)
AR(A)
FI,AF,AT

GARDENS AVIATION, INC.
P.O. Box 12642
Lake Park, FL 33403
AR(A)
FI,AF

GATEWAY AVIATION, INC.
Rt. 2, Box 2A
Titusville, FL 32780
PC(A)
CC(A),IR(A)
FI,AF

GULF ATLANTIC AIRWAYS
4305 NE 49th Dr.
Regional Airport
Gainesville, FL 32609
PC(A)

HOLLYWOOD FLYING SERVICE, INC.
7750 Hollywood Blvd.
Hollywood, FL 33024
PC(A)

ISLAND CITY FLYING SERVICE, INC.
1900 S. Roosevelt Blvd.
Key West, FL 33040
PC(A)
CC(A),IR(A)
FI

JAX NAVY FLYING CLUB
Hangar 116
P.O. Box 127
Jacksonville, FL 31121
PC(A)
CC(A),IR(A)
AR(A),PG(P)(I)(C)(F)(T)(R)
FI,AF,AT

KENDALL FLYING SCHOOL, INC.
P.O. Box 557516
Miami, FL 33155
PC(A)
CC(A),IR(A)
AR(A),PG(P)(I)(C)
FI

L T AERO
14532 SW 129 St.
Miami, FL 33186
PC(A),PT(A)
IR(A)

LAUDERDALE AVIATION, INC.
7501 Pembroke Road
Hollywood, FL 33023
PC(A),PT(A)
CC(A),CT(A),IR(A)
PG(P)(I)(C)(F)(R)
FI,AF

MANATEE JUNIOR COLLEGE
5840 26th St. West
Bradenton, FL 33506
PG(P)(I)(C)

MCARTHUR HIGH SCHOOL
6501 Hollywood Blvd.
Hollywood, FL 33054
PG(P)

MIAMI DADE COMMUNITY COLLEGE
South Campus
11011 SW 104th St.
Miami, FL 33156
PG(P)(I)(C)

MIAMI DADE COMMUNITY COLLEGE NORTH
11380 NW 27th Ave.
Miami, FL 33167
PG(P)

MIAMI HELICOPTER SERVICE, INC.
Building 147 Opa Locka Airport
Opa Locka, FL 33054
AR(R)
FI

MILTON T AVIATION
Milton Flying Service Inc.
P.O. Box 742
Milton, FL 32570
PC(A),PT(A)
CC(A),CT(A),IR(A)
AR(A),PG(P)(I)(C)(F)(T)(R)(A)
FI,AF,AI,AT

MIRACLE STRIP AVIATION, INC.
P.O. Drawer H
Destin, FL 32541
PC(A)
CC(A),IR(A)
AR(A),PG(P)(I)(C)(F)(T)(R)
FI,AT

NATIONAL AVIATION ACADEMY
Delta Aircraft Corp.
St. Petersburg Clearwater Airport
Clearwater, FL 33520
PC(A)
CC(A),IR(A)
AR(A)
FI,AF

PALATKA AVIATION, INC.
Kay Larkin Airport
Rt. 1, Box 1880
Palatka, FL 32077
PC(A)
PG(P)

PELICAN AIRWAYS
World Aircraft Flight Operations Inc.
7501 Pembroke Rd.
N. Perry Airport
W. Hollywood, FL 33023
PC(A),PT(A)
CC(A),CT(A),IR(A)
AR(A)
FI,AF,AI

PENSACOLA AVIATION CENTER, INC.
P.O. Box 2781
Pensacola, FL 32503
PT(A)
CC(A),CT(A),IR(A)
AR(A),PG(P)(I)(C)(F)(T)(R)
FI,AF,AT

PHOENIX EAST AVIATION, INC.
561 Pearl Harbor Dr.
Daytona Beach, FL 32014
PC(A)
CC(A),IR(A)
AR(A),PG(P)(I)(C)(F)(T)(R)
FI,AF,AT

POMPANO AIR CENTER, INC.
1401 NE 10 St.
Pompano Beach, FL 33060
PC(A)
CC(A),IR(A)
AR(A)
FI,AF

QUINCY AVIATION SERVICES, INC.
Quincy Airport
Rt. 6, Box 13A
Quincy, FL 32351
PC(A),PT(A)
CC(A),CT(A),IR(A)
PG(P)(I)(C)

ROBERTS FLYING SERVICE, INC.
P.O. Box 1011
Lakeland, FL 33802
CC(A),IR(A)
AR(A)

SAVCO FLYING
Savco Inc.
7750 Hollywood Blvd.
Hollywood, FL 33024
PC(A)
CC(A),IR(A)
AR(A)

SOUTH DADE AVIATION SERVICES
28700 SW 217 Ave.
Homestead, FL 33030
PC(A)

SOWELL AVIATION COMPANY, INC.
P.O. Box 1490
Panama City, FL 32401
PC(A),PT(A)
CC(A),CT(A),IR(A)
AR(A),PG(P)(I)(C)(F)(T)(R)
FI,AF,AT

ST. PETERSBURG JUNIOR COLLEGE
2465 Drew St.
Clearwater, FL 33515
PG(P)(I)(C)

SUNRAY AIRLINE
1609 Hanger Rd., Building 332
Sanford Airport
Sanford, FL 32771
PC(A)
CC(A),IR(A)
AR(A)
FI,AF,AI

TECHNICAL AVIATION SERVICE, INC.
5553 NW 36 St.
Suite C
Miami Springs, FL 33166
PG(P)(I)(C)

TIDWELL, A. C.
Star Rt. 2, Box 9068
Tallahassee, FL 32304
PC(A)
CC(A),IR(A)
PG(P)(I)(C)(T)

TOPP AIR, INC.
St. Petersburg Clearwater Airport
Clearwater, FL 33520
PC(A)
CC(A),IR(A)
AR(A)
FI,AF

TURSAIR, INC.
Building 147, Opa Locka Airport
P.O. Box 85
Opa Locka, FL 33054
PC(A)
CC(A),IR(A)
AR(A),PG(P)(I)(C)(F)(T)(R)

TYNDALL AFB AERO CLUB
P.O. Box A
Tyndall AFB, FL 32403
PC(A)
CC(A),IR(A)
AR(A)
FI,AF,AT

VENICE FLYING SERVICE, INC.
220 E. Airport Ave.
Venice, FL 33595
PC(A)
CC(A),IR(A)
AR(A)
FI,AF

VOLUSIA AVIATION SERVICE, INC.
Regional Airport
Daytona Beach, FL 32014
PC(A),PT(A)
CC(A),IR(A)
AR(A)
FI,AF,AT

WEST FLORIDA HELICOPTERS, INC.
Albert Whitted Airport, Hangar 1
St. Petersburg, FL 33701
PC(R)
CC(R)
AR(R)
FI

WINGS INTERNATIONAL, INC.
P.O. Box 3288
Ft. Pierce, FL 33454
PC(A)

WORDAIR INTERNATIONAL, INC.
14532 SW 129th St.
Miami, FL 33186
PC(A),PT(A)
PG(P)

GEORGIA

AIR VALDOSTA, INC.
2612 Madison Hwy.
Valdosta, GA 31601
PC(A),PT(A)
CC(A),CT(A),IR(A)
AT

AUGUSTA AVIATION, INC.
Daniel Field
Augusta, GA 30909
PC(A)
CC(A),IR(A)
AR(A)
FI,AF,AT

Butler Aviation Savannah, Inc.
P.O. Box 23647
Savannah, GA 31403
PC(A)
CC(A),IR(A)
AR(A)
FI,AF

Epps Air Service, Inc.
Dekalb Peachtree Airport
Atlanta, GA 30341
PC(A),PT(A)
CC(A),CT(A),IR(A)
AR(A)
FI,AF

Gulfstream Learning Center
Flight International, Inc.
P.O. Box 2307
Travis Field
Savannah, GA 31402
AR(A)

Holland Flying Service, Inc.
Valdosta Municipal Airport
Valdosta, GA 31601
PC(A),PT(A)
CC(A),CT(A)
FI

L and M Aviation, Inc.
14 Airport Rd.
Malcolm McKinnon Airport
St. Simmons Island, GA 31522
PC(A)
CC(A),IR(A)
FI

Peachtree Dakalb Flight Academy, Inc.
1954 Airport Rd.
Dekalb Peachtree Airport
Chamblee, GA 30341
PC(A)
CC(A),IR(A)
AR(A)
FI,AF,AT

Quality Aviation, Inc.
1951 Airport Rd., Suite 102
Dekalb Peachtree Airport
Atlanta, GA 30341
PC(A)
CC(A),IR(A)
AR(A)
FI,AF,AT

Robins Air Force Base Aero Club
Building 184
Robins AFB, GA 31098
PC(A)
CC(A),IR(A)
FI,AF

HAWAII

Associated Aviation Activities
Universal Enterprises Inc.
218 Lagoon Dr.
Honolulu, HI 96819
PC(A)
CC(A),IR(A)
AR(A)
FI,AF

Hawaii Air Academy
3031 Aolele St.
Honolulu, HI 96819
PC(A)
CC(A),IR(A)
AR(A)(G)
FI,AF,AT,PR

James F. Pierce Flight School
C. A. McCluney Co.
404 Aowena Pl.
Honolulu, HI 96819
PC(A)
CC(A),IR(A)
AR(A)
FI,AF,AT

Kona Flight Services
P.O. Box 2067
Kailua-Kona, HI 96740
CC(A),IR(A)
AR(A)
FI,AF,AT

IDAHO

AERO TECHNICIANS, INC.
P.O. Box 7
Rexburg, ID 83440
PC(A)
CC(A)(R)
AR(A)(R),PG(P)(C)
FI,AF

BOISE STATE UNIVERSITY
1910 College Blvd.
Boise, ID 83725
PG(P)(I)(C)

CLARKS AIR SERVICE, INC.
P.O. Box 56
Nampa, ID 83651
PC(A)
CC(A),IR(A)
PG(P)(I)(C)
FI,AF,AG

FLIGHT SERVICE, INC.
P.O. Box 38
Caldwell, ID 83605
PC(A)
CC(A),IR(A)
AR(A),PG(P)(I)(C)(F)(R)
FI,AF

POCATELLO AVCENTER, INC.
17 Star Rt.
Airport
Pocatello, ID 83201
PC(A)

ILLINOIS

AIR INSTITUTE AND SERVICE
Southern Illinois University
Southern Illinois Airport
Carbondale, IL 62901
PC(A)
CC(A),IR(A)
AR(A)
FI

AMCORP
Aviation Management Corp.
P.O. Box 553
Lansing, IL 60438
PC(A)

ASSOCIATED AIR ACTIVITIES
P.O. Box 158
Lansing Municipal Field
Lansing, IL 60438
PC(A)

AVIATION TRAINING ENTERPRISES ATE
American Flyers
Aviation Training Enterprises, Inc.
3N040 Powis Rd.
West Chicago, IL 60185
PC(A)
CC(A),IR(A)
AR(A)
FI,AF,AT,PR

BELLEVILLE AREA COLLEGE
1400 Upper Cahokia Rd.
Cahokia, IL 62206
PC(A)
CC(A),IR(A)
AR(A)
FI,AF

CLARK AVIATION, INC.
Bloomington Normal Airport
Bloomington, IL 61701
PC(A)
CC(A),IR(A)
AR(A)
FI,AF,AT

DIXON AVIATION, INC.
Walgreen Field
Dixon, IL 61021
PC(A),PT(A)
CC(A),CT(A),IR(A)
AR(A)
FI

Dupage Aviation Corp.
Dupage County Airport
West Chicago, IL 60185
PC(A),PT(A)
CC(A),CT(A),IR(A)
AR(A),PG(T)
FI,AF,AT

Elliott Flight Center
Elliott Flying Service, Inc.
Quad City Airport
Box 26
Moline, IL 61265
PC(A),PT(A)
CC(A),CT(A),IR(A)
AR(A)
FI

Galt Flying Service, Inc.
5112 Greenwood Rd.
Wonder Lake, IL 60097
CC(A),CT(A),IR(A)
AR(A)
FI,AF

George J. Priester Aviation Service
Priester, George J.
Pal Waukee Airport
Wheeling, IL 60090
PC(A)
CC(A),IR(A)
AR(A)
FI,AF,AT

Illini Aviation, Inc.
1402 E. Airport Rd.
Frasca Field
Urbana, IL 61801
PC(A)

Parks College of St. Louis University
BI State Parks Airport
Cahokia, IL 62206
PC(A)
CC(A),IR(A)
AR(A)

Prairie State College
197th and Halstead
Chicago Heights, IL 60411
PG(P)

Rock Valley College
6349 Falcon Rd.
Rockford, IL 61109
PG(P)(C)

T K Aviation, Inc.
4943 W. 63rd St.
Chicago, IL 60638
CC(A),IR(A)

Tufts Edgcumbe, Inc.
Elgin Airport
Box 557
Elgin, IL 60120
PC(A),PT(A)
CC(A),CT(A),IR(A)
AR(A)
FI,AF,AT

University of Illinois
Institute of Aviation
Willard Airport Univ. of Illinois
Savoy, IL 61874
PC(A)
CC(A),IR(A)
AR(R)

Vincennes University Aviation
Lawrenceville Vincennes Municipal Airport
Lawrenceville, IL 62439
PC(A)
CC(A),IR(A)

INDIANA

Aretz Flying Service, Inc.
180 Aretz Lane
Lafayette, IN 47905
PC(A)
CC(A),IR(A)
FI,AF

Chambers Aviation, Inc.
Rt. 18, Box 127
Indianapolis, IN 46234
PC(A)
CC(A),IR(A)
FI

FRANKLIN FLYING FIELD, INC.
Rt. 3, Box 58
Franklin, IN 46131
PC(A)
CC(A),CT(A),IR(A)
AR(A)
FI,AF

GRIFFITH AVIATION, INC.
1705 E. Main St.
Griffith, IN 46391
PC(A)
CC(A),IR(A)
FI,AF

H AND D AVIATION, INC.
Hulman Regional Airport
Terre Haute, IN 47803
PC(A),PT(A)
CC(A),CT(A),IR(A)
AR(A)
FI,AF

HELICOPTER AIRWAYS OF INDIANA, INC.
1401 N. Rangeline Rd.
Carmel, IN 46032
PC(R)
CC(R)
AR(R)
FI

INDIANA STATE UNIVERSITY
Dept. of Aerospace Technology
118 N. 6th St.
Terre Haute, IN 47809
PG(P)(I)(C)(F)

INDIANAPOLIS AVIATION, INC.
10565 Allisonville Rd.
Indianapolis, IN 46060
PC(A)
CC(A),IR(A)
FI,AF

LAFAYETTE AVIATION, INC.
Hangar 5, Purdue Univ. Airport
Lafayette, IN 47906
PC(A),PT(A)

PURDUE UNIVERSITY
Dept. of Aviation Technology
Purdue Airport
Lafayette, IN 47906
AR(A)

SKY HARBOR, INC.
7700 W. 38th St.
Indianapolis, IN 46254
PC(A),PT(A)
CC(A),CT(A),IR(A)
AR(A)
FI,AF,AT

SUNRISE AVIATION, INC.
2572 CR NR 60
Auburn, IN 46706
PC(A)(R)
CC(A)(R),IR(A)
AR(A)(R)
FI,AF

TRI STATE AERO, INC.
6101 Flightline Dr.
Evansville, IN 47711
PC(A)

IOWA

BALLOONS OVER IOWA, LTD.
287 190th Ave.
Carlisle, IA 50047
PG(P)

BLUFFS AIRCRAFT, INC.
Municipal Airport
Council Bluffs, IA 51501
PC(A)
CC(A),IR(A)
AR(A)
FI,AF

CHARLES CITY AERONAUTICS, INC.
Rt. 4, Municipal Airport
Charles City, IA 50616
PC(A)
CC(A),IR(A)
AR(A)
FI,AT

DENISON AVIATION, INC.
Municipal Airport
Denison, IA 51442
PC(A)
CC(A),IR(A)

HAPS AIR SERVICE
Haps Air Services, Inc.
Municipal Airport
Ames, IA 50010
PC(A)
CC(A),IR(A)
AR(A)
FI,AF,AT

IOWA CITY FLYING SERVICE, INC.
Municipal Airport
Iowa City, IA 52240
PC(A),PT(A)
CC(A),CT(A),IR(A)
AR(A)
FI,AF,AT

IOWA LAKES COMMUNITY COLLEGE
300 S. 18th St.
Estherville, IA 51334
PC(A)
CC(A),IR(A)

IOWA STATE UNIVERSITY
Ames Municipal Airport
Ames, IA 50010
PC(A)
PG(P)

JOHNSON AVIATION, INC.
Box 986, Municipal Airport
Newton, IA 50208
PC(A)
CC(A),IR(A)
AR(A)
FI,AF

MARSHALLTOWN AVIATION, INC.
Municipal Airport, Rt. 1
Marshalltown, IA 50158
PC(A)
CC(A),IR(A)
AR(A)
FI,AF

P AND N CORP.
McBride Airport
Marion, IA 52302
PC(A),PT(A)
IR(A)

REMMERS TOMKINS FLIGHT SERVICE, INC.
P.O. Box 373, Municipal Airport
Burlington, IA 52601
PC(A)
CC(A),IR(A)
AR(A)
FI,AF,AT

STORM FLYING SERVICE, INC.
Rt. 2, Municipal Airport
Webster City, IA 50595
PC(A),PT(A)
CC(A),CT(A),IR(A)
AR(A)
FI,AF,AT

STRALEY FLYING SERVICE, INC.
Municipal Airport
Davenport, IA 52804
PC(A),PT(A)
CC(A),CT(A),IR(A)
AR(A)
FI,AF,AT

TIBBEN FLIGHT LINES, INC.
Cedar Rapids Municipal Airport
Cedar Rapids, IA 52401
PC(A)
CC(A),IR(A)
AR(A)

UNIVERSITY OF DUBUQUE
Municipal Airport
Dubuque, IA 52001
PC(A)
CC(A),IR(A)
AR(A)
FI,AF

WATHAN FLYING SERVICE, INC.
P.O. Box 1368
Municipal Airport
Cedar Rapids, IA 52406
PC(A)
CC(A),IR(A)
AR(A),PG(P)
FI,AF,AT

KANSAS

CAPITOL AIR SERVICE, INC.
Manhattan Municipal Airport
P.O. Box 763
Manhattan, KS 66502
PC(A)
CC(A),IR(A)
AR(A)
FI,AF,AI,AT

CESSNA EMPLOYEES FLYING CLUB
Cessna Aircraft Co.
1780 Airport Rd.
Mid Continent Airport
Wichita, KS 67209
PC(A),PT(A)
CC(A),CT(A),IR(A)
AR(A)

CITATION LEARNING CENTER
Flightsafety Intl. Inc.
1851 Airport Rd.
P.O. Box 12323
Wichita, KS 67277
AR(A)
PR

COFFEYVILLE AIRCRAFT, INC.
P.O. Box 322
Municipal Airport
Coffeyville, KS 67337
PC(A)
CC(A),CT(A),IR(A)
AR(A)
FI,AF,AT

CRAIGS AERO SERVICE
Craig, Charles H.
Rt. 1, Box 14
Gardner, KS 66030
PC(A)
CC(A),IR(A)
AR(A),PG(P)(I)(C)(F)(T)(R)(A)

FORT RILEY FLYING CLUB
359 Curant Court
Marshall US Army Airfield
Ft. Riley, KS 66442
PC(A),PT(A)
CC(A),CT(A),IR(A)
AR(A)
FI,AF,AT

K.C.H. SCHOOL OF AERONAUTICS
Kings Flying Service, Inc.
One Executive Aero Plaza
Olathe, KS 66062
PC(A)
CC(A)(R),CT(R),IR(A)
AR(A)(R),PG(P)(I)(C)(F)(T)(R)(A)
FI,AF,AT

KANSAS CITY PIPER, INC.
P.O. Box 1850
Olathe, KS 66061
PC(A),PT(A)
CC(A),CT(A),IR(A)
AR(A)
FI,AF,AT

KEN GODFREY AVIATION, INC.
3600 Sardou
Billard Airport
Topeka, KS 66616
PC(A)
CC(A),IR(A)

LAWRENCE AVIATION, INC.
Rt. 3, Municipal Airport
Lawrence, KS 66044
PC(A)
CC(A),IR(A)
PG(P)(I)(C)(F)
FI,AF

MCCONNELL AERO CLUB
United States Air Force
McConnell AFB
Wichita, KS 67221
PC(A),PT(A)
CC(A),CT(A),IR(A)
FI

MOORES MIDWAY AVIATION
Jerry Moores Midway Aviation, Inc.
2812 Hein Ave.
Salina Municipal Airport
Salina, KS 67401
PC(A)

YINGLING AIRCRAFT, INC.
P.O. Box 9248
Mid Continent Airport
Wichita, KS 67209
PC(A)
CC(A),IR(A)
AR(A)
FI,AT

KENTUCKY

AYER FLYING SERVICE, INC.
Owensboro County Airport
P.O. Box 30
Owensboro, KY 42301
CC(A),IR(A)
AR(A),PG(P)(C)
FI,AF

CENTRAL AVIATION, INC.
Madison Airport
Rt. 2, Box 330
Richmond, KY 40475
PC(A)

DON DAVIS AVIATION
Davis, Donald C.
Rt. 2, Box 28
Henderson, KY 42420
PC(A),PT(A)
CC(A),IR(A)
AR(A)

EASTERN KENTUCKY UNIVERSITY
Aviation Program
Stratton Building
Richmond, KY 40475
PG(P)

ELIZABETHTOWN FLYING SERVICE, INC.
Lawson, Roger, E.
Ben Floyd Field, Rt. 8
Elizabethtown, KY 42701
PC(A)
CC(A),CT(A),IR(A)
FI,AF

KENTUCKY FLYING SERVICE, INC.
Bowman Field, Building T30
Louisville, KY 40205
PC(A)
CC(A),IR(A)
AR(A)
FI,AF,AT

LOUISIANA

AERO FLYING SERVICE
Angel Aviation, Inc.
231 Airport Rd.
Slidell, LA 70458
PC(A)
CC(A),IR(A)
AR(A)
FI

AIRTAIX AVIATION, INC.
Williams Mitchell Hangars
New Orleans Lakefront Airport
New Orleans, LA 70126
PC(A)
CC(A),IR(A)
AR(A)
FI,AF

DERIDDER AIRCRAFT CORP.
2132 Blankenship Dr.
Deridder, LA 70634
PC(A),PT(A)
CC(A),CT(A),IR(A)
AR(A)
FI,AF

FLEEMAN AVIATION
Fleeman Enterprises, Inc.
5410 Operations Rd.
Monroe, LA 71201
PC(A)
CC(A),CT(A),IR(A)
AR(A)
FI,AF

LAKEFRONT FLIGHT SPECIALTIES, INC.
P.O. Box 26574
New Orleans, LA 70126
PC(A)
CC(A),IR(A)
AR(A)
FI,AF,AT

LINCOLN SERVICES, INC.
S. Farmerville St.
Ruston Airport
Ruston, LA 71270
PC(A),PT(A)
CC(A),IR(A)

LOUISIANA TECH UNIVERSITY
Dept. of Professional Aviation
P.O. Box 6445, Tech Station
Ruston, LA 71270
PC(A)
CC(A),IR(A)
AR(A),PG(P)(I)(C)
FI,AF

LOUISIANA TECH UNIVERSITY, BARKSDALE
P.O. Box 128
Barksdale, LA 71010
PG(P)(I)(C)(F)(T)(R)(A)

NEW ORLEANS AVIATION, INC.
Walter Wedell Hangar
Lakefront Airport
New Orleans, LA 70126
PC(A)
CC(A),IR(A)
FI,AF

NICHOLLS AVIATION, INC.
P.O. Box 1222
Thibodaux, LA 70302
PC(A)
CC(A),IR(A)
AR(A)

NORTHWESTERN STATE UNIVERSITY
Natchitoches Parish Airport
Natchitoches, LA 71457
PC(A)
CC(A),IR(A)
AR(A),PG(P)(I)(C)(F)(T)
FI,AT

PELICAN AVIATION CORP.
P.O. Box 2008
Acadiana Regional Airport
New Iberia, LA 70560
PC(A),PT(A)
CC(A),IR(A)
AR(A),PG(P)
FI,AF,AI

SOUTHERN AVIATION CORP.
1400 Airport Dr.
Shreveport, LA 71107
PC(A)
CC(A),IR(A)
AR(A)

TIGER AIR CENTER, INC.
5520 Operations Rd.
Monroe, LA 71203
PC(A),PT(A)

MAINE

CENTRAL MAINE FLYING SERVICE, INC.
DeWitt Field
Old Town, ME 94468
PC(A),PT(A)
CC(A),CT(A),IR(A)
AR(A)
FI,AF

MARYLAND

AERO FLIGHT LTD.
7940 Airpark Dr.
Gaithersburg, MD 20879
PC(A)
CC(A),IR(A)
AR(A)
FI,AF,AT

ATC FLIGHT TRAINING CENTER
Computer Flight Instruction, Inc.
6709 Cpl. Frank Scott Dr.
College Park, MD 20740
PG(P)(I)(C)

BALTIMORE AVIATION SERVICE, INC.
Baltimore Airpark
White Marsh, MD 21162
PC(A),PT(A)
CC(A),CT(A),IR(A)
PG(P)(I)(C)
FI,AF

CUMBERLAND AIRLINES
Nicholson Air Services, Inc.
P.O. Box 1611
Cumberland Municipal Airport
Cumberland, MD 21502
PC(A)
CC(A),IR(A)
PG(P)(I)(C)

EASTERN FLYING SERVICE, INC.
Essex Skypark
1401 Diffendall Rd.
Baltimore, MD 21221
PC(A)

FREDERICK AVIATION, INC.
Frederick Municipal Airport
Frederick, MD 21701
PC(A)
CC(A)

FREEWAY AIRPORT, INC.
3900 Church Rd.
Mitchellville, MD 20716
PC(A)
CC(A),IR(A)
FI,AF

GIBSON AVIATION, INC.
7940 Airpark Dr.
Gaithersburg, MD 20760
PC(A),PT(A)
CC(A),CT(A),IR(A)
AR(A)
FI,AF

HINSON AIRWAYS, INC.
P.O. Box 8709
BWI Airport
Baltimore, MD 21240
PC(A)
CC(A),CT(A),IR(A)
AR(A)
FI,AF

MARYLAND AIRLINES CO., INC.
P.O. Box 577
Municipal Airport
Easton, MD 21601
PC(A)
CC(A),IR(A)
FI,AF

MCDONALD FLYING SERVICE
McDonald, Billy S.
St. Mary's County Airport
Star Rt., Box 315
California, MD 20619
PC(A)
CC(A),IR(A)
PG(P)(I)(C)(F)
FI

PROFESSIONAL FLIGHT SERVICE
9550 Allentown Rd.
Oxon Hill, MD 20022
PC(A)
CC(A),IR(A)
AR(A),PG(P)(I)(C)(T)
FI,AF,AT

SUBURBAN AIRSERVICE, INC.
520 Brock Bridge Rd.
Laurel, MD 20810
PC(A),PT(A)
CC(A),CT(A),IR(A)
AR(A)
FI,AF,AT

MASSACHUSETTS

AMITY AEROTECH INSTITUTE
Worcester Municipal Airport
Worcester, MA 01602
PC(A),PT(A)
CC(A),IR(A)

AVIATION EAST, INC.
Vaccaro, William F.
Norfolk Airport
River Rd.
Norfolk, MA 02056
AF

AVIATION TRAINING ACADEMY
Aviation Training Academy, Inc.
Sterling Airport
Sterling Junction, MA 01565
PC(A)(R)
CC(A)(R),IR(A)
AR(R)

BRIDGEWATER STATE COLLEGE
Science Building
Bridgewater, MA 02324
PG(P)(I)(C)(F)(A)

CITY AVIATION, INC.
LaFleur Airport
P.O. Box 221
Northhampton, MA 01060
CC(A),CT(A),IR(A)

COMERFORD WEST, INC.
Worcester Municipal Airport
Worcester, MA 01602
PC(A)
CC(A),IR(A)
AR(A)

E.W. WIGGINS AIRWAYS, INC.
Norwood Municipal Airport
Norwood, MA 02062
PC(A)
CC(A),IR(A)
AR(A)(R),PG(P)(I)(C)
FI,AF,AI

EXECUTIVE FLYERS AVIATION
Goulian, Myron
Hanscom Field
Bedford, MA 01730
PC(A)
CC(A),IR(A)
AR(A)
FI,AF

FITCHBURG COLONIAL AVIATION, INC.
Fitchburg Municipal Airport
Fitchburg, MA 01420
PC(A),PT(A)
CC(A),CT(A),IR(A)
AR(A)
FI,AF

HYANNIS AVIATION, INC.
Barnstable Municipal Airport
Hyannis, MA 02601
PC(A)
CC(A),IR(A)

KING SERVICES OF AVIATION, INC.
Taunton Municipal Airport
East Taunton, MA 02718
PC(A)
CC(A),IR(A)
PG(P)(I)(C)
FI,AF

NEW ENGLAND FLYERS AIR SERVICE, INC.
Beverly Airport
Beverly, MA 01915
PC(A)
CC(A),IR(A)
AR(A),PG(P)(I)(C)(F)(T)(R)
FI,AF,AI,AT

NORTH SHORE COMMUNITY COLLEGE
3 Essex St.
Beverly, MA 01915
PG(P)(I)(C)

PATRIOT AVIATION CORP.
State Terminal Building
Hanscom Field
Bedford, MA 01730
PC(A)

ROBERTS AVIATION
Matukaitis, Wallace R.
Metro Airport
Palmer, MA 01069
PC(A)
CC(A),IR(A)
FI

SOUTH WEYMOUTH NAVAL AERO CLUB
US Naval Air Station
South Weymouth, MA 02190
PC(A)
CC(A),IR(A)
FI,AF

Tew Mac School of Aeronautics
Main St.
Tewksbury, MA 01876
PC(A)

Yankee Aviation
Air Charter Express, Inc.
Plymouth Municipal Airport
Plymouth, MA 02360
PC(A),PT(A)
CC(A),CT(A),IR(A)
AR(A), PG(P)(I)(C)(F)(R)
FI, AF

MICHIGAN

American Wings and Wheels, Inc.
P.O. Box 7275
Flint, MI 48507
PC(A)
CC(A),IR(A)
FI

Andrews University Aviation
Andrews Airport
Berrien Springs, MI 49104
PC(A)
CC(A),IR(A)
AR(A)
FI,AF

Anglin Flying Service
G3101 W. Bristol Rd.
Flint, MI 48507
PC(A)
IR(A)

Ann Arbor Aero Service
Hinshaw Aviation, Inc.
4320 State Rd.
Ann Arbor, MI 48104
PC(A)

Anna's Airways, Inc.
1675 Airport Rd.
Oakland Pontiac Airport
Pontiac, MI 48054
PC(A),PT(A)
CC(A),CT(A),IR(A)
AR(A)
FI

Aviation Group, Inc.
1525 Airport Rd.
Pontiac, MI 48054
PC(A),PT(A)
CC(A),CT(A),IR(A)

Aviation Services, Inc.
5580 Airport Rd.
Mt. Pleasant, MI 48858
PC(A)

Brooks Aero, Inc.
1243 S. Kalamazoo
Marshall, MI 49068
PC(A)

D J's Flying Service
D J's Aviation, Inc.
48000 Tyler Rd.
Belleville, MI 48111
PC(A)
CC(A),IR(A)
PG(P)(I)(C)

Drake Aviation Co., Inc.
Oakland Pontiac Airport
6330 Highland Rd.
Pontiac, MI 48054
PC(A),PT(A)
CC(A),CT(A),IR(A)
AR(A)
FI,AF,AT

Flight One, Inc.
1510 Main
Owosso, MI 48867
PC(A)
IR(A)

Flytel Aviation
Exec. Terminal Suite 114
Detroit City Airport
Detroit, MI 48213
PC(A)

Grand Rapids School of the Bible and Music
1331 Franklin SE
Grand Rapids, MI 49506
PC(A)
CC(A),IR(A)
AR(A)(R),PG(F)(R)

GREAT LAKES AERO
1232 Roods Rd.
Lapeer, MI 48446
PC(A)
CC(A),IR(A)

GROSSE ILE FLIGHT SERVICES, INC.
9505 Groh Rd.
Grosse Ile, MI 48138
PC(A)
CC(A),IR(A)
AR(A)
FI,AF

HENRY FORD COMMUNITY COLLEGE
5101 Evergreen Rd.
Dearborn, MI 48128
PG(P)

JACKSON COMMUNITY COLLEGE
3610 Wildwood Ave.
Jackson, MI 49202
PC(A)
IR(A)

JACOBS FLYING SERVICE, INC.
Lenawee County Airport
2651 Cadmus Rd.
Adrian, MI 49221
PC(A),PT(A)

JET SERVICES, INC.
Mettetal Airport
8850 Lilley Rd.
Canton, MI 48187
PC(A)

KAL AERO, INC.
5605 Portage Rd.
Kalamazoo, MI 49002
PC(A),PT(A)
CC(A),CT(A),IR(A)
AR(A)
FI,AF

LANSING COMMUNITY COLLEGE
419 N. Capital Ave.
Box 40010
Lansing, MI 48901
PC(A)
CC(A),IR(A)
PG(P)(I)(C)

MACOMB AVIATION CORP.
59819 Indian Trail Rd.
New Haven, MI 48048
PC(A)
CC(A),IR(A)

MCKINLEY EXECUTIVE CORP.
1851 W. Maple Rd.
Troy, MI 48084
PC(A),PT(A)
CC(A),CT(A),IR(A)

MERILLAT SCHOOL OF AVIATION
5447 Rogers Hwy.
Tecumseh, MI 49286
PC(A)

MICHIGAN AERO CORP.
12401 Conner Ave.
Detroit, MI 48205
PC(A)
CC(A),IR(A)

NORTHERN AIR SCHOOL OF AERONAUTICS
Northern Air Service, Inc.
5500 44th St.
Grand Rapids, MI 49508
PC(A),PT(A)
CC(A),IR(A)
AR(A),PG(I)(T)
FI,AF,AT

OAKLAND COMMUNITY COLLEGE
2480 Opdyke Rd.
Bloomfield Hills, MI 48013
PG(P)(I)(C)

OLSEN FLIGHT SERVICE, INC.
11499 Conner Rd., Room 114
Detroit, MI 48213
PC(A)
CC(A),IR(A)
AR(A),PG(P)(I)(C)(F)(R)
FI,AF,AT

PRENTICE AIRCRAFT, INC.
2495 Cadmus Rd.
Lenawee County Airport
Adrian, MI 49221
PC(A),PT(A)

PRICE AVIATION
618 Silver Lake Rd.
Linden, MI 48451
PC(A)
CC(A)

R J HELICOPTERS, INC.
51125 Poniact Trail
Wixom, MI 48096
PC(R)
CC(R)
AR(R)
FI,AF

ROISEN ENTERPRISES, INC.
719 Airport Dr.
Ann Arbor, MI 48104
PC(A)

SCHOOLCRAFT COLLEGE
18600 Haggerty Rd.
Livonia, MI 48152
PG(P)(I)(C)

SHAL AERO
9505 Groh Rd.
Grosse Ile, MI 48138
PC(A)

SKYBOLT AVIATION
G 3101 W. Bristol Rd.
Flint, MI 48507
IR(A)

SPICER FLYING SERVICE
Executive Terminal, Suite 103
11201 Corner
Detroit, MI 48213
PC(A)

TRADEWINDS AVIATION, INC.
6545 Highland Rd.
Pontiac, MI 48054
PC(A),PT(A)
CC(A),CT(A)
AR(A)

WESTERN MICHIGAN UNIVERSITY
Kalamazoo County Airport
Kilgore Rd.
Kalamazoo, MI 49001
CC(A)

MINNESOTA

AERODROME INC.
Rochester Municipal Airport
Rochester, MN 55901
PC(A)
CC(A),IR(A)
AR(A)
FI,AF,AT

CALL O WILD AIR
Kranz, Roger W.
Rt. 5, Mankato Airport
Mankato, MN 56601
PC(A)
CC(A),IR(A)
AR(A)
AF

CIRRUS FLIGHT OPERATIONS, INC.
2289 County Rd. J
Blaine, MN 55432
PC(A),PG(P)
CC(A)
FI,AF

CRYSTAL SHAMROCK INC.
6000 Douglas Dr. North
Minneapolis, MN 55429
PC(A)
CC(A),IR(A)
AR(A),PG(P)
FI,AF,AT

FLIGHT TRAINING CENTER, INC.
9960 Flying Cloud Dr.
Eden Prairie, MN 55343
PC(A)(R),PT(A)
CC(A)(R),CT(A),IR(A)
AR(A)(R)

FLYING SCOTCHMAN
Arneson, Roy B.
6300 Zane Ave. North
Brooklyn Park, MN 55429
PC(A)
CC(A)
AR(A)
FI,AF

INSTRUMENT FLIGHT TRAINING
Instrument Flight Training of Minn.
590 Bayfield St.
St. Paul Downtown Airport
St. Paul, MN 55107
PC(A)
CC(A),IR(A)
AR(A),PG(P)(I)(C)
FI,AF,AT

SKYLINE FLITE, INC.
2289 County Rd. J
Minneapolis, MN 55432
PC(A)
CC(A),IR(A)
AR(A)
FI,AF,AT

ST. CLOUD AVIATION, INC.
P.O. Box 1599
St. Cloud, MN 56302
PC(A)

ST. CLOUD STATE UNIVERSITY
Dept. of Technology
St. Cloud, MN 56301
PG(P)(I)(C)

SUBURBAN HENNEPIN COUNTY VO TECH
Independent Schl. Dist. 287
9200 Flying Cloud Dr.
Eden Prairie, MN 55344
PG(P)(I)(C)

THUNDERBIRD AVIATION, INC.
14091 Pioneer Trail
Eden Prairie, MN 55343
PC(A)
CC(A),IR(A)
AR(A)
FI,AF,AT

THUNDERBIRD AVIATION OF CRYSTAL, INC.
Crystal Airport
Crystal, MN 55429
PC(A)
CC(A),CT(A),IR(A)
AR(A)
FI,AF,AI

UNIVERSITY OF MINNESOTA
Flight Facilities
2289 County Rd. J
Minneapolis, MN 55432
PC(A)
CC(A),IR(A)
AR(A),PG(P)
FI,AF,AT

UNIVERSITY OF MINNESOTA
Technical College
Hwys. 2 and 75
Crookston, MN 56716
PC(A)
CC(A)

WINGS, INC.
St. Paul Downtown Airport
St. Paul, MN 55107
PC(A)
CC(A),IR(A)
AR(A)
FI,AF,AT

WINONA STATE UNIVERSITY
General Delivery
Winona, MN 55987
PG(P)

MISSISSIPPI

AIR COMMAND
Etheridge, Dorothy K.
P.O. Box 5421
Greenville, MS 38701
PC(A)
CC(A),IR(A)

COTTON BELT AVIATION, INC.
Rt. 1, Box 482
Greenwood, MS 38930
PC(A),PT(A)
CC(A),CT(A),IR(A)

MERIGOLD FLYING SERVICE, INC.
P.O. Box 307
Merigold, MS 38759
CC(A)
AG

MISSISSIPPI SCHOOL OF AVIATION, INC.
556 West Ramp
Jackson, MS 39209
PC(A),PT(A)
CC(A),CT(A),IR(A)
AR(A)
FI,AF

MISSISSIPPI STATE UNIVERSITY
Drawer A
Mississippi, MS 39762
PG(P)

VICTORY AIRCRAFT, INC.
Jackson County Airport
P.O. Box 340
Pascagoula, MS 39567
PC(A)
CC(A),IR(A)
FI,AF

MISSOURI

ARCHWAY AVIATION, INC.
3127 Creve Coeur Mill Rd.
St. Louis, MO 63141
PC(A)
CC(A),IR(A)
FI,AF

B J'S PILOT GROUND SCHOOL
B J's Enterprises, Inc.
250 Richards Rd., Suite 266N
Kansas City, MO 64116
PG(P)(I)(C)

CAPE CENTRAL AIRWAYS, INC.
Cape Girardeau Municipal Airport
P.O. Box 99
Cape Girardeau, MO 63701

CENTRAL MISSOURI STATE UNIVERSITY
Warrensburg
Warrensburg, MO 64093
PC(A),PT(A)
CC(A),CT(A),IR(A)
PG(P)(I)(C)

DICK HILL HELICOPTERS AND FLYING SERVICE, INC.
Airpark South, Rt. 2
Ozark, MO 65721
PC(R)
CC(R),CT(R)
AR(R)
FI,AF,AG

FLIGHTSAFETY INTERNATIONAL, INC.
6161 Aviation Dr.
St. Louis, MO 63134
AR(A),PG(T)
AT,PR

HAL AVIATION, INC.
1701 E. 5th St.
Kennett, MO 63857
PC(A)(R)
CC(A)(R),IR(A)
AR(A)
FI,AF,AT,AG

MERAMEC COMMUNITY COLLEGE
11333 Big Bend Blvd.
St. Louis, MO 63122
PG(P)(I)(C)

METROPOLITAN HELICOPTERS, INC.
Rt. 1, Box 521A
Glencoe, MO 63038
AR(R)
FI

MIZZOU AVIATION
R L S Rental Co., Inc.
Municipal Airport
Joplin, MO 64802
PC(A),PT(A)
CC(A),CT(A),IR(A)
AR(A),PG(F)(R)
FI,AF,AT

PENN VALLEY COMMUNITY COLLEGE
3201 SW Trafficway
Kansas City, MO 64111
PG(P)(I)(C)

RANKIN AIRCRAFT
Rankin, Joe
Rankin Airport, Rt. 3
Maryville, MO 64468
PC(A),PT(A)
PG(P)

ROEDERER AVIATION, INC.
Spirit of St. Louis Airport
525 Mercury Blvd.
Chesterfield, MO 63017
PC(A)
CC(A),IR(A)
FI,AF

SCHOOL OF THE OZARKS, INC.
General Delivery
Point Lookout, MO 65726
PC(A)
CC(A),IR(A)
PG(P)
FI,AT

ST. CHARLES FLYING SERVICE
3001 Airport Rd.
St. Charles, MO 63301
PC(A)
CC(A)(G),IR(A)
AR(A),PG(P)(I)(C)
FI,AF,AT

TRANS WORLD AIRLINES, INC.
1307 Baltimore
Kansas City, MO 64105
AR(A)
AT

MONTANA

DILLON FLYING SERVICE
Morris, James Andre
Box 188
Dillon, MT 59725
PC(A)
CC(A),IR(A)
FI,AF

LYNCH FLYING SERVICE, INC.
Logan Intl. Airport
Billings, MT 59105
PC(A)
CC(A),IR(A)
AR(A)
FI,AF,AT

MILES CITY AERO SERVICE, INC.
P.O. Box 656
Miles City, MT 59301
PC(A),PT(A)
CC(A),CT(A)
AR(A)
FI

SUNBIRD AVIATION, INC.
P.O. Box 808
Belgrade, MT 59714
PC(A),PT(A)
CC(A),CT(A),IR(A)
AR(A)

VALLEY AVIATION
Rt. 4032
Great Falls Intl. Airport
Great Falls, MT 59401
PC(A)
CC(A),IR(A)

NEBRASKA

AIRKAMAN OF OMAHA, INC.
P.O. Box 19064
Eppley Airfield
Omaha, NE 68119
PC(A),PT(A)
CC(A),CT(A),IR(A)
AR(A)
FI,AF,AT

LINAIRE, INC.
Lincoln Municipal Airport
Lincoln, NE 68524
PC(A)

LINCOLN AVIATION INSTITUTE, INC.
Municipal Airport
Lincoln, NE 68524
PC(A),PT(A)
CC(A),CT(A),IR(A)
AR(A)(R)
FI,AT

MIDWAY AVIATION, INC.
P.O. Box 1844
Kearney, NE 68847
PC(A),PT(A)
CC(A),CT(A),IR(A)

MIDWEST AIRWAYS, INC.
Rt. 1, Municipal Airport
Plattsmouth, NE 68048
PC(A),PT(A)
CC(A),CT(A),IR(A)
AR(A)
FI,AF

OFFUTT AERO CLUB
P.O. Box 13234
Offutt AFB, NE 68113
PC(A),PT(A)
CC(A),CT(A),IR(A)
AR(A)
FI,AF,AT

SCOTTSBLUFF AVIATION CO., INC.
Rt. 2, Box 182J
Scottsbluff, NE 69361
PC(A),PT(A)
CC(A),CT(A),IR(A)
AR(A)
FI,AF,AT

SKY CRAFT, INC.
P.O. Box 477
Grand Island, NE 68801
PC(A)
CC(A),IR(A)

SOUTHEAST COMMUNITY COLLEGE
8800 O St.
Lincoln, NE 68520
PG(A)

TREGO AVIATION, INC.
P.O. Box 1226
North Platte, NE 69101
PC(A),PT(A)
CC(A),CT(A),IR(A)
AR(A)
FI,AF

NEVADA

AERLEON
2772 N. Rancho Dr.
Las Vegas, NV 89130
PC(A),PT(A)
CC(A),IR(A)
FI,AF

AIR NEVA, INC.
Cannon Intl. Airport
2601 E. Plumb Lane
Reno, NV 89502
PC(A),PT(A)
CC(A),CT(A),IR(A)
AR(A)
FI,AT

AVIATION SERVICES, INC.
1880 Gentry Way
Reno, NV 89502
PC(A),PT(A)
CC(A),CT(A),IR(A)
FI

CARSON TAHOE AVIATION, INC.
2600 E. Graves Lane
Carson City, NV 89701
PC(A)
CC(A),IR(A)
AR(A)
FI,AF

GENERAL AVIATION SERVICES
5425 Texas Ave.
Reno, NV 89506
PC(A)
CC(A),IR(A)
FI,AF

NEVADA AERO FLIGHT ACADEMY
Alta Sierra Aviation
2500 Graves Lane
Carson City, NV 89701
PC(A),PT(A)
CC(A),CT(A),IR(A)
AR(A)
FI,AF

NEVADA AVIATION SERVICES, INC.
2772 Rancho Drive
Las Vegas, NV 89106
PC(A),PT(A)
CC(A),IR(A)
AR(A)
FI

WINNEMUCCA AIR SERVICE
Municipal Airport
Winnemucca, NV 89445
AR(A)
FI

NEW HAMPSHIRE

ALLEN FLYING SERVICE, INC.
238 Rochester Hill Rd.
Rochester, NH 03867
PC(A),PT(A)
PG(P)

DANIEL WEBSTER COLLEGE
University Dr.
Nashua, NH 03063
PC(A)(G)
CC(A)(G),IR(A)
PG(P)(I)(C)

FERNS FLYING SERVICE, INC.
Concord Municipal Airport
Concord, NH 03301
PC(A)
CC(A),IR(A)
PG(P)(I)(C)(F)
FI,AF

LEBANON AIRPORT DEVELOPMENT CORP.
Lebanon Regional Airport
West Lebanon, NH 03784
PC(A),PT(A)
CC(A),IR(A)
AR(A)

NASHUA AVIATION AND SUPPLY CO., INC.
Nashua Municipal Airport
Boire Field
Nashua, NH 03060
PC(A)
CC(A),IR(A)
AR(A)
FI,AF,AT

NATHANIEL HAWTHORNE COLLEGE
Hawthorne-Feather Airpark
Antrim, NH 03440
PC(A)
CC(A),IR(A)

NEW JERSEY

CHEROKEE AERO CLUB
Aero Venture, Inc.
106B Sharon Rd.
Robbinsville, NJ 08691
PC(A),PT(A)
CC(A),CT(A),IR(A)
AR(A),PG(P)
FI,AF

MCGUIRE AFB AERO CLUB
P.O. Box 838
Wrightstown, NJ 08562
PC(A)
CC(A),IR(A)
FI,AF

MERCER COUNTY COMMUNITY COLLEGE
1200 Old Trenton Rd.
Trenton, NJ 08690
PC(A)
CC(A),IR(A)
PG(P)(I)(C)

RARITAN VALLEY FLYING SCHOOL
Princeton Airport
Rt. 206
Princeton, NJ 08540
PC(A),PT(A)
CC(A),CT(A),IR(A)
AR(A)
FI,AF,AT

SANDPIPER AIR SERVICES
Building 101
Millville Airport
Millville, NJ 08332
PC(A)
CC(A),IR(A)
AR(A)

SOLBERG FLIGHT TRAINING CENTER
Solberg Airport
P.O. Box 250
Somerville, NJ 08876
PC(A)
CC(A),IR(A)
AR(A)

NEW MEXICO

DOUBLE EAGLE AVIATION, INC.
P.O. Box 10310
Albuquerque, NM 87184
PC(A),PT(A)
CC(A),CT(A),IR(A)
AR(A)
FI,AF,AT

ED'S FLYING SERVICE, INC.
Alamogordo White Sands
Regional Airport, Box 966
Alamogordo, NM 88310
PC(A)
CC(A),IR(A)
AR(A),PG(P)(I)(C)(F)(T)(R)
FI,AF,AT

GREAT SOUTHWEST AVIATION, INC.
P.O. Box 5700
Roswell, NM 88201
PC(A)
CC(A),IR(A)

J M GRIMES AVIATION
Silver City Grant County Airport
P.O. Box 2119
Silver City, NM 88062
PC(A)
CC(A),IR(A)
AR(A),PG(P)(I)(C)

KIRTLAND AFB AERO CLUB
Hangar 333
Albuquerque, NM 87117
PC(A)
CC(A),IR(A)
AR(A)
FI,AF,AT

MESA AVIATION SERVICES, INC.
P.O. Box 1105
Farmington, NM 87401
PC(A)
CC(A),IR(A)
FI,AF

SEVEN BAR FLYING SERVICE, INC.
10001 Coors Rd., NW
Albuquerque, NM 87114
PC(A)

SOUTHWEST AVIATION, INC.
P.O. Box 387
Fairacres, NM 88033
PC(A)
CC(A),IR(A)
AR(A),PG(P)(I)(C)(F)(T)(R)
FI,AF,AT

WORLD BALLOON CORP.
4800 Eubank NE
Albuquerque, NM 87111
PC(L)
CC(L)
AR(L)

NEW YORK

ACADEMICS OF FLIGHT
43-49 45th St.
Sunnyside, NY 11104
CC(A),IR(A)
PG(P)(I)(C)(F)(T)(R)

AIR EXPERTS
2111 Smithtown Ave.
Ronkonkoma, NY 11779
IR(A)
AR(A),PG(I)(F)(T)(R)
FI,AF,AT

ATE OF NEW YORK, INC.
Gen. Avia. Terminal
Long Island MacArthur Airport
Ronkonkoma, NY 11779
PC(A)
CC(A),IR(A)
AR(A)
FI,AF,AT,PR

AVIATION HIGH SCHOOL
Queens Blvd. and 36th St.
Long Island City, NY 11101
PG(P)

BOARD OF COOPERATIVE EDUCATIONAL SERVICES
610 Hicksville Rd.
Bethpage, NY 11714
PG(P)

BOARD OF COOPERATIVE EDUCATIONAL SERVICES
375 Locust Ave.
Oakdale, NY 11769
PG(P)

BROCKWAY AIR INC., NY
Clinton County Airport
Plattsburgh, NY 12901
PC(A)
CC(A),CT(A),IR(A)
AR(A),PG(P)(I)(C)
FI,AF

BUFFALO AIRFIELD MANAGEMENT CORP.
4500 Clinton St.
Buffalo, NY 14224
PC(A)
CC(A),IR(A)
FI,AF

DOWLING COLLEGE
Idle Hour Blvd.
Oakdale, NY 11769
PG(P)(I)(C)(F)

DUNKIRK AVIATION FLIGHT SCHOOL
Dunkirk Municipal Airport
Dunkirk, NY 14048
PC(A)
CC(A),IR(A)
AR(A)
FI,AF,AT

EAST HAMPTON FLIGHT SERVICES, INC.
P.O. Box 656
Wainscott, NY 11975
PC(A)
CC(A),IR(A)
AR(A)
FI,AF

EAST HILL FLYING CLUB, INC.
Tompkins County Airport
Ithaca, NY 14580
PC(A)
CC(A),CT(A),IR(A)
FI,AF

EDISON TECHNICAL AND OCCUPATIONAL EDUCATION CENTER
655 Colfax St.
Rochester, NY 14606
PC(A)

GACE FLYING CLUB, INC.
P.O. Box 335
Bethpage, NY 11714
PG(P)(I)(C)

GREECE CENTRAL SCHOOL DISTRICT
Continuing Education Div.
P.O. Box 300
North Greece, NY 14515
PG(P)(I)

GREENLAND SCHOOL OF AVIATION
Greenland Air, Inc.
1 Mill Rd.
Latham, NY 12110
PC(A),PT(A)
CC(A),CT(A),IR(A)
AR(A),PG(P)
FI,AT

ISLAND HELICOPTERS, INC.
Island Heliport North Ave.
Garden City, NY 11530
PC(R)
CC(R),CT(R),IR(H)
AR(R)
FI,AF,AI,AT

MID COUNTY FLYERS, INC.
2099 Smithtown Ave.
Long Island MacArthur Airport
Ronkonkoma, NY 11779
PC(A)
CC(A),IR(A)
PG(P)(I)(C)

MID ISLAND FLYING SCHOOL
Mid Island Air Service, Inc.
Long Island MacArthur Airport
Ronkonkoma, NY 11779
PC(A)
CC(A),IR(A)
AR(A)
FI,AF,AT

MILLER AVIATION, INC.
P.O. Box 564
Endicott, NY 13760
PC(A)
CC(A),IR(A)
AR(A)
FI,AF,AT

PAGE ROCHESTER BEECHCRAFT AERO CENTER
Page Avjet Corp.
1265 Scottsville Rd.
Rochester, NY 14624
PC(A)
CC(A),CT(A),IR(A)
AR(A)

PAN AMERICAN AVIATION INSTITUTE, INC.
74-09 37th Ave.
Jackson Heights, NY 11372
PG(P)(I)(C)(F)(T)

PERRY'S FLYING SERVICE, INC.
Building 128
Suffolk County Airport
Westhampton, NY 11978
PC(A)
CC(A),IR(A)

QUEENSBOROUGH COMMUNITY COLLEGE
56th Ave. & Springfield Blvd.
Bayside, NY 11364
PG(P)(I)

RENSSELAER LEARNING SYSTEMS
One Rensselaer Dr.
Pittsford, NY 14534
PG(P)(I)

RICHMOR AVIATION, INC.
Columbia County Airport
Hudson, NY 12534
PC(A),PT,(A)
CC(A),IR(A)
FI,AF,AT

SAIR AVIATION FLIGHT SCHOOL
1801 Malden Rd.
Syracuse, NY 13211
PC(A),PT(A)
CC(A),CT(A),IR(A)
AR(A)
FI,AF,AT

SCHWEIZER SOARING SCHOOL
P.O. Box 147
Elmira, NY 14902
PC(G)
AR(G)

SKY LIFE FOUNDATION, INC.
P.O. Box 699
Northville, NY 12134
PC(A),PT,(A)
CC(A),CT(A),IR(A)
PG(P)(I)(C)

WELLSVILLE FLYING SERVICE
P.O. Box 641
Wellsville, NY 14895
PC(A),PT(A)
CC(A),CT(A)
AR(A)
FI,AF,AT

WESTAIR FLYING SCHOOL
Westchester Aeronautical Corp.
Westchester County Airport
White Plains, NY 10604
PC(A)
CC(A),IR(A)
AR(A)
FI,AF,AI,AT

NORTH CAROLINA

ALLISON AVIATION
Allison, Michael S.
Rt. 4, Box 43, Homestead MHP
Greenville, NC 27834
PC(A)

ATLANTIC AERO, INC.
P.O. Box 19608
Greensboro, NC 27419
PC(A)
CC(A),IR(A)
AR(A)
FI,AF,AT

BALLOON ASCENSIONS, LTD.
Rt. 11, Box 97
Statesville, NC 28677
PC(L)
CC(L)

CAROLINA AIR ACADEMY, INC.
Elkin Airport
Elkin, NC 28621
PG(P)(I)(C)

CHERRY POINT MARINE AERO CLUB
P.O. Box 918
Havelock, NC 28532
PC(A)
CC(A),IR(A)
AR(A)
FI,AF

CONNON AVIATION
Carolina Airways, Inc.
P.O. Box 1968
Hickory, NC 28601
PC(A)
CC(A),IR(A)
FI,AF

GOLDSBORO WAYNE AVIATION, INC.
P.O. Box 386
Pikeville, NC 27863
PC(A)
CC(A),IR(A)
FI,AF

GUILFORD TECHNICAL COMMUNITY COLLEGE
P.O. Box 309
Jamestown, NC 27282
PG(P)(I)(C)

ISO AERO SERVICE, INC.
P.O. Box 1294
Kinston, NC 28501
PC(A),PT(A)
CC(A),CT(A),IR(A)
AR(A)
FI,AF

LENOIR COMMUNITY COLLEGE
P.O. Box 188
Kinston, NC 28501
PG(P)(I)(C)

MISSIONARY AVIATION
Piedmont Bible College, Inc.
Sugar Valley Airport
Rt. 2, Box 381
Mocksville, NC 27028
PC(A)
CC(A),IR(A)

SEYMOUR JOHNSON AERO CLUB
Seymour Johnson AFB, NC 27530
PG(F)(R)
PC(A)
CC(A),IR(A)

SHELBY AVIATION, INC.
P.O. Box 970
Shelby, NC 28150
PC(A)
CC(A),IR(A)
AR(A)

TAR HEEL AVIATION, INC.
P.O. Box 911
Jacksonville, NC 28540
PC(A)
CC(A),IR(A)

WAYNE COMMUNITY COLLEGE
Caller Box 8002
Goldsboro, NC 27530
PG(P)

NORTH DAKOTA

UNIVERSITY OF NORTH DAKOTA
Dept. of Aviation
Box 8216, Univ. Station
Grand Forks, ND 58201
PC(A)(R),PT(A)
CC(A)(R),CT(A),IR(A)
AR(A)
FI,AF,AT

OHIO

AIR SERVICE CENTER, INC.
2820 Bobmeyer Rd.
Hamilton, OH 45015
PC(A),PT(A)
CC(A),CT(A),IR(A)
AR(A)
FI,AF,AT

ALLEN COUNTY AVIATION CORP.
700 Airport Dr.
Lima, OH 45804
PC(A)
CC(A),IR(A)
AR(A)
FI,AF,AT

AM AIR
Schuster, George
Youngstown Municipal Airport
P.O. Box 205
Vienna, OH 44473
PC(A),PT(A)
CC(A),CT(A),IR(A)
AR(A)
FI,AF

AVIATION SALES, INC.
10600 Springboro Pike
Miamisburg, OH 45342
PC(A)
CC(A),IR(A)
FI,AF

AVIATION SALES, INC.
591 N. Dixie Dr.
Vandalia, OH 45377
PC(A)
CC(A),IR(A)
FI,AF

BOWLING GREEN STATE UNIVERSITY
School of Technology
Bowling Green, OH 43403
PC(A)
CC(A),IR(A)

BROOKVILLE AIR PARK, INC.
9386 National Rd.
Brookville, OH 45309
PC(A)
CC(A),IR(A)
AR(A)
FI,AF,AT

CARDINAL AIR TRAINING, INC.
Terminal Building, Lunken Airport
Cincinnati, OH 45226
PC(A)(R)
CC(A)(R),IR(A)
AR(A)(R)
FI,AF,AT

CAS AVIATION INC. FLITE CENTER
1954 Norton Rd.
Columbus, OH 43228
PC(A),PT(A)
IR(A)

CENTRAL SKYPORT, INC.
4700 E. Fifth Ave.
Columbus, OH 43219
PC(A),PT(A)
CC(A),IR(A)

CHOI AVIATION FLIGHT SCHOOLS
Choi, George Y.H.
1800 Triplett Blvd.
Akron, OH 44306
PC(A),PT(A)
CC(A),CT(A),IR(A)
AR(A)

CHOI AVIATION, INC.
Wadsworth Municipal Airport
900 Airport Dr.
Wadsworth, OH 44281
PC(A),PT(A)
CC(A),CT(A),IR(A)
FI,AF

CITATION LEARNING CENTER
Flight Safety Intl., Inc.
Rt. 4, Box 295C
Swanton, OH 43558
AR(A)
AT,PR

COMMONWEALTH AVIATION, INC.
P.O. Box 75052
Greater Cincinnati Airport
Cincinnati, OH 45275
PC(A)

COMMUNITY AIRPORT CORP.
2050 Medina Rd.
Medina, OH 44256
PC(A)
CC(A),IR(A)

CUYAHOGA COMMUNITY COLLEGE
Western Campus
11000 W. Pleasant Valley Rd.
Parma, OH 44130
PG(P)(I)(C)(F)

EXECUTIVE JET AVIATION, INC.
625 N. Hamilton Rd.
Columbus, OH 43219
AR(A)
PR

GALION FLIGHT TRAINING, INC.
8240 State Rt. 309
Galion Municipal Airport
Galion, OH 44833
PC(A)
CC(A),IR(A)
AR(A)

HEYDE AVIATION CENTER, INC.
10-646 County Road O
Napoleon, OH 43545
PC(A)
CC(A),IR(A)
AR(A)
FI,AF

ICARUS EXECUTIVE SERVICES, INC.
523½ N. Columbus St.
Suite D
Lancaster, OH 43130
PC(A)
CC(A)

KENT STATE UNIVERSITY
Aerospace Technology
4020 Kent Rd.
Stow, OH 44224
PC(A)
CC(A),IR(A)
PG(P)(I)(C)

KETTERING ADULT SCHOOL
3700 Far Hills Ave.
Kettering, OH 45429
PG(P)

LAWRENCE COUNTY AVIATION, INC.
Rt. 2, Box 369
South Point, OH 45680
PC(A)
CC(A),IR(A)
PG(P)

LIGHTNING AVIATION, INC.
Sidney Airport
14833 Sidney Plattsville Rd.
Sidney, OH 45365
PC(A),PT(A)
CC(A),CT(A),IR(A)

LORAIN COUNTY COMMUNITY COLLEGE
1005 N. Abbe Rd.
Elyria, OH 44035
PG(P)

MADISON AVIATION CENTER, INC.
P.O. Box 287
London, OH 43140
PC(A)
CC(A),IR(A)
AR(A)
FI,AF

MIAMI UNIVERSITY
Miami University Airport
Fairfield Rd.
Oxford, OH 45056
PC(A)
CC(A),IR(A)

OHIO UNIVERSITY
Ohio University Airport
Athens, OH 45701
PC(A)
IR(A)
AR(A)
FI,AF

PERSONAL FLIGHT TRAINING, INC.
270 Wilmer Rd.
Lunken Airport
Cincinnati, OH 45226
PC(A)

STAFFORD FLIGHT ACADEMY
Wings Inc.
1800 Triplett Blvd.
Akron, OH 44306
PC(A),PT(A)
CC(A),CT(A),IR(A)
AR(A)
FI,AF

SUNDORPH AERONAUTICAL CORP.
Flight School
Cleveland Hopkins Intl. Airport
Cleveland, OH 44135
PC(A)
CC(A),IR(A)
FI,AF

THE OHIO STATE UNIVERSITY
Box 3022
University Station
Columbus, OH 43210
PC(A)
CC(A),IR(A)
AR(A),PG(P)
FI,AF,AT,PR

WRIGHT PATTERSON AFB AERO CLUB
Building 153, Area C
Wright Patterson AFB, OH 45433
PC(A)
CC(A),IR(A)
AR(A)
FI,AF,AT

WRIGHT STATE UNIVERSITY
3640 Colonel Glenn Hwy.
Dayton, OH 45435
PG(P)

5 K FLIGHTS, INC.
Hangar W 10
Cleveland Hopkins Intl. Airport
Cleveland, OH 44135
PC(A)
CC(A),IR(A)
PG(P)(I)(C)
FI,AF

OKLAHOMA

AIRCRAFT SERVICE CO.
Grosh, Donald R.
Harvey Young Airport
1500 S. 135th E Ave.
Tulsa, OK 74108
PC(A)
CC(A)(R),IR(A)
AR(A)(R)
FI,AF

ALLIED HELICOPTER SERVICE, INC.
P.O. Box 6216
Tulsa Downtown Airport
Tulsa, OK 74148
AR(R)
AF,AG,XL

DELTA AVIATION
Delacerda, Fred
1802 W. Wright
Stillwater, OK 74075
PC(A)

EXEC EXPRESS, INC.
2020-4 W. Airport Rd.
Stillwater, OK 74075
PC(A),PT(A)
CC(A),CT(A),IR(A)
AR(A)

MIDLAND FLYERS, INC.
Miami Municipal Airport
P.O. Box 905
Miami, OK 74354
PC(A)
CC(A),IR(A)
AR(A)

NATIONAL EDUCATION CENTER
National Education Centers, Inc.
Spartan School of Aeronautics Campus
P.O. Box 51133
Tulsa, OK 74132
PC(A)
CC(A),IR(A)
AR(A)
FI,AF,AT

OKLAHOMA HELICOPTERS, INC.
1700 Lexington
Room 108
Norman, OK 73069
PC(R)
CC(R)
FI,AF,XL

OKLAHOMA STATE UNIVERSITY
Aviation Education Dept.
414 Classroom Building OSU
Stillwater, OK 74078
PG(P)(I)(C)

PIONAIR FLYING CLUB
Western Oklahoma State College
2801 N. Main
Altus, OK 73521
PC(A)
PG(P)(I)(C)(F)(T)

PORT CHEROKEE SEAPLANE BASE, INC.
Rt. 2, Box 826, Grand Lake
Afton, OK 74331
AR(A)

REDLEG FLYING CLUB
Henry Post Airfield
P.O. Box 33307, Building 4914
Post Rd.
Ft. Sill, OK 73503
PC(A)
CC(A),IR(A)
AR(A),PG(P)(I)(C)(F)(R)

SELLERS AVIATION CO.
Hunter, Robert L.
P.O. Box 204
Enid, OK 73702
PC(A)
CC(A),IR(A)
AR(A),PG(P)(I)(C)(F)(T)(R)(A)
FI,AF,AI,AT,PR

STEVENSON AVIATION SERVICE, INC.
P.O. Box 1711
Davis Field
Muskogee, OK 74401
PC(A),PT(A)
CC(A),CT(A),IR(A)

STILLWATER FLIGHT CENTER, INC.
2020-3 W. Airport Rd.
Stillwater, OK 74075
PC(A)
CC(A),IR(A)
PG(P)(I)(C)

UNIVERSITY OF OKLAHOMA
Aviation Dept.
1700 Lexington Dr.
Norman, OK 73069
PC(A)
CC(A),IR(A)
PG(P)(I)(C)

VERSATILE HELICOPTERS, INC.
P.O. Box 1433
Downtown Ardmore Airport
Ardmore, OK 73401
PC(A)
CC(R)
AR(R)

OREGON

AAR WESTERN SKYWAYS, INC.
Portland Troutdale Airport
Troutdale, OR 97060
PC(A),PT(A)
CC(A),CT(A),IR(A)
AR(A)
FI,AF,AT

AURORA AVIATION
Northwest Aircraft Rental, Inc.
22775 Airport Rd. NE
P.O. Box 127
Aurora, OR 97002
PC(A)
CC(A),IR(A)
AR(A)
FI,AF,AT

CASCADE FLIGHT CENTER
Mobley, Donald E.
Dalles Municipal Airport
P.O. Box 771
Dallesport, OR 98617
PC(A)
FI

CAVEMAN AVIATION, INC.
2280 Carton Way
Grants Pass, OR 97526
PC(A)
CC(A),IR(A)
AR(A)
FI,AF,AI

COOS AVIATION, INC.
1210 Airport Way
North Bend, OR 97459

EAGLE FLIGHT CENTER, INC.
Portland Hillsboro Airport
Hillsboro, OR 97123
PC(A)
CC(A),IR(A)
AR(A)
FI,AF,AT

EUGENE FLIGHT CENTER, INC.
28809 Airport Rd.
Eugene, OR 97402
PC(A)
CC(A),IR(A)
AR(A)
FI,AF

HERMISTON FLIGHT CENTER
Langdon, Wrex
P.O. Box 814
Hermiston, OR 97838
PC(A)
CC(A),IR(A)
FI

HILLSBORO HELICOPTERS, INC.
3301A NE Cornell Rd.
Hillsboro, OR 97124
PC(A)(R)
CC(A)(R),IR(A)
AR(R)

HORIZON AVIATION, INC.
P.O. Box 509
Aurora, OR 97002
PC(A)
CC(A),IR(A)
AR(A)
FI,AF

KLAMATH AIRCRAFT, INC.
Municipal Airport
Oregon City, OR 97601
CC(A),IR(A)
AR(A)
FI,AF

LANE COMMUNITY COLLEGE
28715 Airport Rd.
Eugene, OR 97402
PC(A)
CC(A),IR(A)
AR(A),PG(P)(I)(C)(F)
FI,AF

LOGAN AND REAVIS AIR, INC.
Medford Jackson County Airport
Medford, OR 97504
PC(A)
CC(A),CT(A),IR(A)
FI,AF

MCKENZIE FLYING SERVICE, INC.
90600 Greenhill Rd.
Eugene, OR 97402
PC(A)
CC(A),IR(A)
AR(A),PG(T)
FI,AF

MOUNTAIN AIR HELICOPTERS, INC.
3510 Knox Butte Rd.
Albany, OR 97321
PC(A)(R)
CC(A)(R)
AR(A)(R)
FI,AF,AI,AT,XL

ONTARIO FLIGHT SERVICE
Rt. 3, Box 11
Ontario, OR 97914
PC(A),PT(A)
CC(A),CT(A),IR(A)
AR(A),PG(P)(I)(C)(F)(T)(R)
FI

PILOT PERSONNEL INTERNATIONAL, INC.
1600 Airway Dr.
Lebanon, OR 97355
IR(A)
AR(A)
FI,AT,XL

ROSEBURG SKYWAYS, INC.
Associated Capital Corp.
Municipal Airport
Roseburg, OR 97470
PC(A)
CC(A),IR(A)
AR(A)
FI,AF

SALEM AVIATION, INC.
2680 Aerial Way SE
Salem, OR 97302
PC(A)
CC(A),IR(A)
AR(A)
FI,AF

SOUTHERN OREGON SKYWAYS, INC.
401 Dead Indian Rd.
Ashland, OR 97520
PC(A),PT(A)
CC(A),CT(A),IR(A)
AR(A)
FI,AF,AI,AT

SOUTHERN OREGON STATE COLLEGE
1250 Siskiyou Blvd.
Ashland, OR 97520
PG(P)

STONE AVIATION, INC.
3443 NE Cornell Rd.
Hillsboro, OR 97124
PC(A)

TRANSWESTERN HELICOPTERS, INC.
P.O. Box R
Scappoose, OR 97056
PC(R)
CC(R)
AR(R)
FI,XL

TREASURE VALLEY COMMUNITY COLLEGE
650 College Blvd.
Ontario, OR 97914
PG(P)(I)(C)(F)

PENNSYLVANIA

AERO FLIGHT, INC.
Harlan, Charles
Rt. 2, Box 292
Montoursville, PA 17754
PC(A)
CC(A),IR(A)

AGROTORS INC.
Box 578
Gettysburg, PA 17325
PC(R)
CC(R)
AR(R)
FI,AF,AT,AG,XL

AIR CONDOR HELICOPTERS
Air Condor, Inc.
609 Ross Ave.
New Cumberland, PA 17070
PC(R)
CC(R),CT(R)
AR(R)
FI,AF,AT

ARNER FLYING SERVICE, INC.
Jake Arner Airport
Lehighton, PA 18235
PC(A)
CC(A),IR(A)
AR(A)
FI

BEAVER AVIATION SERVICE, INC.
Beaver County Airport
Beaver Falls, PA 15010
PC(A),PT(A)
CC(A),CT(A),IR(A)
AR(A)
FI,AF,AT

BRADENS FLYING SERVICE, INC.
3800 Sullivan Trail
Easton, PA 18042
PC(A)
CC(A),IR(A)
AR(A)
FI,AF,AT

CAP AVIATION, INC.
P.O. Box 13037
Reading Municipal Airport
Reading, PA 19612
PC(A)
CC(A),IR(A)
PG(P)(I)(C)

CLARK AVIATION CORP.
201 Airport Rd.
Capital City Airport
New Cumberland, PA 17070
PC(A)
CC(A),IR(A)
AR(A)
FI,AF,AI,AT

COMMUNITY COLLEGE OF BEAVER COUNTY
College Dr.
Monaca, PA 15061
PG(P)(I)(C)

COMMUNITY COLLEGE OF ALLEGHENY COUNTY
Allegheny Campus
Avia. Bus. Dept. M513
808 Ridge Ave.
Pittsburgh, PA 15212
PG(P)(C)

EAST STROUDSBURG STATE COLLEGE
Physics Dept.
East Stroudsburg, PA 18301
PG(P)

FLYING DUTCHMAN AIR SERVICE, INC.
Buehl Field
Langhorne, PA 19047
PC(A)
CC(A),IR(A)

HASKI AVIATION
New Castle Airport
New Castle, PA 16101
PC(A),PT(A)
CC(A),CT(A),IR(A)
FI,AF

HORSHAM VALLEY AIRWAYS, INC.
451 Caredean Dr.
Horsham, PA 19044
PC(R)
CC(R)

LANCASTER AVIATION, INC.
Rt. 3
Lititz, PA 17543
PC(A)
CC(A)(R),IR(A)
AR(A)(R)
FI,AF,AT

LINCOLN FLYING SERVICE
North Philadelphia Airport
Granted Ashton Rds.
Philadelphia, PA 19114
PC(A)

METRO AIR, INC.
Allegheny County Airport
West Mifflin, PA 15122
PC(A)
CC(A),IR(A)

MOORE AVIATION SERVICE
Beaver County Airport
Beaver Falls, PA 15010
PC(A)
CC(A),IR(A)
AR(A)
FI,AF

MOYER AVIATION, INC.
Box 275
Mount Pocono, PA 18344
PC(A)
CC(A),IR(A)

NEW GARDEN FLYING FIELD
New Garden Aviation, Inc.
P.O. Box 171
Toughkenamon, PA 19374
PC(A),PT(A)
CC(A),CT(A),IR(A)
PG(P)(I)(C)
FI,AF

PENNSYLVANIA STATE UNIVERSITY
2535 Fox Hill Rd.
University Park, PA 16802
PC(A)
CC(A),IR(A)
PG(P)
FI,AF

PENNSYLVANIA STATE UNIVERSITY
Berks Campus
Rt. 5, Box 2150, Tulpehocken Rd.
Reading, PA 19608
PG(P)(I)
PR

QUEEN CITY AVIATION, INC.
1730 Vultee St.
Queen City Airport
Allentown, PA 18103
PC(A),PT(A)
CC(A),CT(A),IR(A)
AR(A)
FI,AF

Security Airways, Inc.
Rostraver Airport
Belle Vernon, PA 15012
PC(A)
CC(A),IR(A)

Steel City Aviation
Allegheny County Airport
West Mifflin, PA 15122
PC(A)
CC(A),IR(A)

Stensin Aviation, Inc.
Beaver County Airport
Beaver Falls, PA 15010
PC(A)
CC(A),CT(A),IR(A)
FI,AF

Summit Airlines, Inc.
Scott Plaza II
Philadelphia, PA 19113
AR(A)

Swarthmore College
Dept. of Physics
Swarthmore, PA 19081
PG(P)

Turner Field, Inc.
1435 Horsham Rd.
Ambler, PA 19002
PC(A),PT(A)
CC(A),CT(A),IR(A)
AR(A)
FI,AF,AT

USAir, Inc.
Greater Pittsburgh Intl. Airport
Pittsburgh, PA 15231
AR(A)

Vee Neal, Inc.
Rt. 1, Box 397
Westmoreland County Airport
Latrobe, PA 15650
PC(A),PT(A)

Wings, Inc.
Rt. 6, P.O. Box 6420
Mercer, PA 16137
PC(A)

RHODE ISLAND

Helms Westerly Aviation, Inc.
Westerly State Airport
Westerly, RI 02891
PC(A),PT(A)
CC(A),IR(A)

North Central Airways, Inc.
North Central State Airport
Lincoln, RI 02865
PC(A)
CC(A),IR(A)
AR(A)
FI,AF,AT

Skylanes, Inc.
North Central State Airport
Lincoln, RI 02865
PC(A),PT(A)
CC(A),CT(A),IR(A)
AR(A)
FI,AF,AT,PR

SOUTH CAROLINA

Eagle Aviation, Inc.
2861 Aviation Way
West Columbia, SC 29169
PC(A)
CC(A),IR(A)
AR(A)
FI,AF,AI,AT

Midlands Aviation Corp.
P.O. Box 5775
Columbia, SC 29250
PC(A)
CC(A),IR(A)
AR(A),PG(P)(I)(C)(F)(T)(R)
FI,AF,AT

North American Institute of Aviation of South Carolina, Inc.
P.O. Box 680
Conway Horry County Airport
Conway, SC 29526
PC(A),PT(A)
CC(A),CT(A),IR(A)
AR(A),PG(P)(I)(C)(F)(T)(R)
FI,AF,AT

ORRCO CORP.
P.O. Box 1023
Spartanburg, SC 29304
PC(A)
CC(A),IR(A)
AR(A),PG(P)(I)(C)(F)(T)(R)
FI,AF,AT

SHAW AFB AERO CLUB
P.O. Box 86
Shaw AFB, SC 29152
PC(A)
CC(A),IR(A)
PG(P)(I)(C)(F)(R)
FI,AF

SKYWAY AVIATION FLIGHT TRAINING
Rt. 3, Box 7
Walterboro, SC 29488
PC(A)
CC(A),IR(A)
PG(P)(I)(C)

SOUTH CAROLINA HELICOPTERS, INC.
P.O. Box 636
Saluda, SC 29138
PC(A)
CC(A)
AR(R),PG(F)
FI,AF,AT,PR,AG

SUMMERVILLE AVIATION, INC.
101 Airport Entrance
Summerville, SC 29483
PC(A)
CC(A),IR(A)
AR(A),PG(P)(I)(C)(F)(T)(R)
FI,AF

SOUTH DAKOTA

AUGUSTANA COLLEGE
29 & S. Summit
Sioux Falls, SD 57102
PG(P)(I)(C)

B AND L AVIATION, INC.
Rt. 2, Box 4590
Rapid City, SD 57701
PC(A)
CC(A),IR(A)
AR(A)
FI,AF,AT

BUSINESS AVIATION, INC.
3501 Aviation Ave.
Sioux Falls, SD 57104
PC(A)
CC(A),IR(A)
AR(A)
FI,AF

FALCON AVIATION, INC.
Rt. 2, Box 297
Yankton, SD 57078
PC(A)
CC(A),IR(A)

JENSEN FLYING SERVICE
Brookings Municipal Airport
Rt. 2
Brookings, SD 57006
PC(A)

MILLER AVIATION
Miller, Keith E.
Brookings Municipal Airport
Rt. 2
Brookings, SD 57006
PC(A)

PROFESSIONAL FLIGHT SERVICES
Hybertson, H.L.
3701 N. Minnesota Ave.
Joe Foss Field
Sioux Falls, SD 57104
PC(A)
CC(A),IR(A)
AR(A)
FI,AF

SILVER WINGS AVIATION, INC.
P.O. Box 522
Rapid City, SD 57709
PC(A),PT(A)
CC(A),CT(A),IR(A)
AR(A)
FI,AF

STAR AVIATION
Star Inc.
Rt. 2, Box 396A
Spearfish, SD 57783
PG(P)

TENNESSEE

BOLIVAR AVIATION
Bolivar Aircraft Sales
P.O. Box 376
Memphis, TN 38008
PC(A),PT(A)
CC(A),CT(A),IR(A)
AR(A),PG(P)(I)(C)(F)(R)
FI,AF

FLIGHT SERVICES, INC.
2787 N. Second St.
Memphis, TN 38127
PC(A),PT(A)
IR(A)

MID TENN. AIRCRAFT, INC.
Rt. 1, Box 698B
Columbia, TN 38401
PC(A)
CC(A),IR(A)

MOODY AVIATION
Moody Bible Institute of Chicago
Box 429
Elizabethton, TN 37643
PC(A)
CC(A),CT(A)

RED AERO, INC.
2488 Winchester, Room 307
Memphis, TN 38116
PC(A)
CC(A),IR(A)
AR(A)
FI,AF

SIGNAL AVIATION SERVICES, INC.
P.O. Box 22100
Chattanooga, TN 37422
PC(A)

SMYRNA AIR CENTER, INC.
Hangar 621 Smyrna Municipal Airport
Smyrna, TN 37167
PC(A)

STEVENS BEECH AERO CLUB
P.O. Box 17248
Nashville Metro Airport
Nashville, TN 37217
PC(A),PT(A)
CC(A),CT(A),IR(A)
AR(A)
FI,AF

VOLUNTEER AVIATION OF KNOXVILLE, INC.
P.O. Box 804
Alcoa, TN 37701
PC(A)
CC(A)

VOLUNTEER FLIGHT TRAINING, INC.
200 Airport Rd.
Clarksville, TN 37040
PC(A)
CC(A),IR(A)
AR(A)
FI,AF

WILLIAMS AVIATION, INC.
P.O. Box 397
Springfield Airport
Springfield, TN 37172
PC(A),PT(A)
CC(A),CT(A),IR(A)
AR(A)
FI,AF,AI

TEXAS

A T FESS MOORE GROUND SCHOOL
8605 Lemmon Ave.
Dallas, TX 75209
PG(P)(I)(T)

ABILENE AERO, INC.
Rt. 2, Box 508
Abilene, TX 79601
PC(A)
CC(A),IR(A)
AR(A)
FI,AF,AT

ACME SCHOOL OF AERONAUTICS
Martonak, Stephen H.
Terminal Building, Meacham Field
Fort Worth, TX 76106
IR(A)

ADDISON PIPER FLITE CENTER
Addison Air Charters, Inc.
4790 Airport Pkwy.
Addison, TX 75001
PC(A),PT(A)
CC(A),IR(A)
AR(A)

AERO ACADEMY, INC.
8244 Travelair
Houston, TX 77061
PC(A),PT(A)
CC(A),CT(A),IR(A)
PG(P)(I)(C)

AERO DYNAMICS, INC.
Dynamic Enterprises, Inc.
8036 Aviation Pl., Nr. 56
Dallas, TX 75235
PC(A),PT(A)
CC(A),CT(A),IR(A)
AR(A)
FI,AF,AI

AEROCOUNTRY AVIATION, INC.
Rt. 1, Rockhill Rd.
Aerocountry Airport
McKinney, TX 75069
PC(A),PT(A)

AIR CENTRAL, INC.
Harlingen Industrial Airport
P.O. Drawer 2623
Harlingen, TX 78550
PC(A)
CC(A),IR(A)
AR(A)(R)
FI,AF

AIRPORT FLYING SCHOOL
Ivy Aviation, Inc.
4511 Eddie Rickenbacker Rd.
Dallas, TX 75248
PC(A)
CC(A),IR(A)
AR(A)
FI,AF,AT

ALPHA AVIATION, INC.
8629 Lemmon Ave.
Dallas, TX 75209
PC(A)
CC(A),IR(A)
AR(A)

ALPHA TANGO FLYING SERVICE
Taylor, Alyce S.
1110 99th Ave.
San Antonio, TX 78214
PC(A),PT(A)
CC(A),CT(A),IR(A)
FI,AF,AT

AMARILLO FLYING SERVICE
Brown, Kenneth V.
Rt. 7, Box 5
Amarillo, TX 79118
CC(A),IR(A)
AR(A),PG(P)
FI,AF,AT

AMERICAN AIR ACADEMY
Rt. 1, Denton Municipal Airport
Denton, TX 76205
PC(A)
CC(A),IR(A)
AR(A)

AMERICAN AIRLINES TRAINING CORP.
American Airlines Flight Academy
P.O. Box 619615
DFW Airport, TX 75261
AR(A)
AT,PR

AMERICAN FLYERS
Location 26 North
Meacham Field
Ft. Worth, TX 76106
PC(A)
CC(A),IR(A)
AR(A)
FI,AF,AI

ANGEL AVIATION, INC.
12510 S. Green Dr., Suite B01
Houston, TX 77034
PG(P)(I)(C)

ANGLO AMERICAN INTL. FLIGHT ACADEMY
500 NW 38th St., Hangar 30S
Meacham Field
Ft. Worth, TX 76106
PC(A)
CC(A),IR(A)
AR(A)
FI

AUSTIN BUSINESS JETS
Corporate Wings, Inc.
1801 E. 51st St.
Austin, TX 78723
PC(A)
CC(A),IR(A)
AR(A)
FI,AF,AT

AVIATION GROUND SCHOOL
Miller, Jack L.
Room B18, Meacham Field Terminal
Ft. Worth, TX 76106
PC(A),PT(A)
CC(A),CT(A),IR(A)
PG(P)(I)(C)(F)(T)

AVIATION TRAINING ENTERPRISES OF TEXAS
7515 Lemmon Dr.
Dallas, TX 75209
PC(A)
CC(A),IR(A)
AR(A)
FI,AF,AT,PR

BAY AREA AVIATION, INC.
Johnson, R.O.
RWJ Airpark
Baytown, TX 77520
PC(A),PT(A)
IR(A)
PG(P)(I)

BELL HELICOPTER TRAINING SCHOOL
Bell Helicopter Textron
Trinity Blvd. and Norwood Dr.
P.O. Box 482
Fort Worth, TX 76101
PC(R)
PR

BROWNING AERIAL SERVICE, INC.
Robert Mueller Municipal Airport
P.O. Box 609
Austin, TX 78767
PC(A),PT(A)
CC(A),CT(A),IR(A)
AR(A)
FI,AF

CENTRAL TEXAS COLLEGE
Hwy. 190 West
Killeen, TX 76541
PC(A)
CC(A),IR(A)
AR(A),PG(P)
FI,AF,AI,AT

CHAPARRAL AVIATION, INC.
4451 Glenn Curtis Dr.
Addison, TX 75001
PC(A)
CC(A),IR(A)
AR(A)

CLIFF HYDE FLYING SERVICE, INC.
11015 W. Main
Laporte, TX 77571
PC(A),PT(A)
CC(A),CT(A),IR(A)
FI,AF

COOK'S FLYING SERVICE
Cook, Floyd W.
Box 606
Iowa Park, TX 76367
PC(A),PT(A)

COTHRON AVIATION, INC.
5104 S. Collins, Rt. 3
Arlington, TX 76014
PC(A)
CC(A),IR(A)
AR(A),PG(P)(I)(C)(F)(T)(R)
FI,AF,AI,AT

D AND J FLIGHT TRAINING
4514 Jacksboro Hwy.
Wichita Falls, TX 76302
PC(A)

DALLAS PIPER
Aero Express, Inc.
5201 S. Hampton Rd.
Lock Box 0-5
Dallas, TX 75232
PC(A),PT(A)
CC(A),CT(A),IR(A)
AR(A)
FI,AF

DOC'S AVIATION SERVICE
Smith, Clark L., Jr.
8650 Cardinal Lane
Smithfield, TX 76180
PC(A),PT(A)

EASTEX AVIATION, INC.
Gregg County Airport
Longview, TX 75601
PC(A)
CC(A),IR(A)
AR(A)
FI,AF

EXECUTIVE AIRCRAFT SERVICES OF TEXAS, INC.
Easterwood Airport
P.O. Box 2828
College Station, TX 77841
PC(A)

FLETCHER AVIATION
8904 Randolph
Houston, TX 77061
PC(A),PT(A)
CC(A),CT(A),IR(A)
PG(T)

FLIGHT PROFICIENCY SERVICE, INC.
P.O. Box 7510
Dallas, TX 75209
PC(A)
CC(A),IR(A)
AR(A)
FI,AF,AT,PR

GENE'S FLIGHT CLUB OF EL PASO, INC.
6805 Boeing
El Paso, TX 79925
PC(A)
CC(A),IR(A)
AR(A), PG(P)(I)(C)(F)
FI,AF

GOBLE AVIATION
Goble, Ted M.
Redbird Airport
Dallas, TX 75232
IR(A)
AR(A)

GRAYSON FLYING SERVICE, INC.
P.O. Box 668
Sherman, TX 75090
PC(A)
CC(A),IR(A)
AR(A)
FI,AF

HANK'S FLITE CENTER, INC.
Box 6036
Midland, TX 79701
PC(A)
CC(A),IR(A)
AR(A),PG(P)(I)(C)(T)
FI,AF,AT

HELI DYNE TRAINING CENTER
Heli Dyne Systems, Inc.
9000 Trinity Blvd.
Ft. Worth, TX 76118
PC(R)
CC(R),IR(H)
AR(R)
FI,AF,AI,AT

HUNT SCHOOL OF AVIATION
H and R Leasing, Inc.
Brownsville Intl. Airport
Brownsville, TX 78521
PC(A),PT(A)
CC(A),CT(A),IR(A)
AR(A)
FI,AF

JET EAST, INC.
7363 Cedar Springs
Dallas, TX 75235
AR(A)

JET FLEET CORP.
7515 Lemmon Ave.
Dallas, TX 75209
IR(A)
AR(A)
AT

LETOURNEAU COLLEGE
P.O. Box 7001
Longview, TX 75607
PC(A)
CC(A),IR(A)

MAJORS AVIATION SERVICES, INC.
P.O. Box 1907
Greenville Municipal Airport
Greenville, TX 75401
PC(A)

MCALLEN AVIATION
Pike Enterprises, Inc.
2812 S. 10th St.
McAllen, TX 78504
PC(A),PT(A)
CC(A),CT(A),IR(A)

MCCREERY AVIATION CO., INC.
P.O. Box 1659
Miller Intl. Airport
McAllen, TX 78502
PC(A)
CC(A),IR(A)

MCKINNEY AVIATION, INC.
McKinney Municipal Airport
P.O. Box 719
McKinney, TX 75069
PC(A)
CC(A),IR(A)

MODERN AERO OF TEXAS
Redbird Airport, Lock Box 37
Dallas, TX 75232
PC(A)
CC(A),IR(A)
AR(A)
FI,AF

MOUNTAIN VIEW COLLEGE
4849 W. Illinois Ave.
Dallas, TX 75211
PG(P)(I)(C)

NAVARRO COLLEGE
P.O. Box 1170
Corsicana, TX 75110
PC(A)
CC(A),IR(A)
PG(P)(I)(C)(F)(T)(R)

NORTH DALLAS HELICOPTER CORP.
4576 Claire Chennault, Suite 12
Dallas, TX 75248
PC(R)
CC(R)
AR(R)
FI

NORTH LAKE COLLEGE
2000 Walnut Hill Lane
Irving, TX 75062
PG(P)(I)(C)

OAKGROVE FLYING SCHOOL
Oakgrove Airport
Fr. Rd. 1187E
Fort Worth, TX 76140
PC(A)
CC(A),IR(A)

PEGASUS FLIGHT CENTER, INC.
Suite 239, Terminal Building
Meacham Field
Fort Worth, TX 76106
PC(A)
CC(A),IR(A)
AR(A)
FI,AF,AT

PLANE THING, INC.
Rt. 4, Box 150
Killeen, TX 76541
PC(A),PT(A)
CC(A),CT(A),IR(A)
AR(A)
FI,AF

PRECISION FLIGHT
Memmi, Linsey
P.O. Box 57
Addison Airport
Addison, TX 75001
PC(A)
CC(A),IR(A)
AR(A),PG(P)(I)(C)
FI,AF

QUALIFLIGHT TRAINING, INC.
Location 1N Meacham Field
P.O. Box 4424
Fort Worth, TX 76106
PC(A)
CC(A),IR(A)
AR(A),PG(P)(I)(C)(F)(R)
FI,AF,AT

RANDOLPH AFB AERO CLUB
P.O. Box 494
Randolph AFB, TX 78148
PC(A)
CC(A),IR(A)
AR(A)
FI,AF,AT,PR

REGIONAL AVIATION
3114 S. Great Southwest Pkwy.
Grand Prairie, TX 75051
PC(A),PT(A)

SAN JACINTO COLLEGE
8060 Spencer Hwy.
Pasadena, TX 77505
PG(P)(I)(C)

SKY BREEZE AVIATION, INC.
Rt. 3, Box 50C
Lubbock Intl. Airport
Lubbock, TX 79401
PC(A)
CC(A),IR(A)

SKYWINGS FLIGHT TRAINING
Yammine, Walid J.
Room 119, Meacham Field Terminal
Fort Worth, TX 76106
PC(A)
CC(A),IR(A)
AR(A)

SOUTHWEST AIR CENTER
Rollins Enterprises, Inc.
2750 FM 1266
League City, TX 77573
PC(A)
CC(A),IR(A)
AR(A),PG(P)(I)(C)

SOUTHWEST TEXAS JUNIOR COLLEGE
Garner Field Rd.
Uvalde, TX 78801
PC(A),PT(A)
CC(A),CT(A),IR(A)
FI

STRAMEL AVIATION, INC.
P.O. Box 863178
Plano, TX 75086
PC(A)
CC(A),IR(A)
AR(A)
FI,AF

TEXARKANA COMMUNITY COLLEGE
2500 N. Robison Rd.
Texarkana, TX 75501
PC(A)
CC(A),IR(A)
PG(P)(I)(C)

TEXAS AERO, INC.
Madison Cooper Airport
P.O. Box 5337
Waco, TX 76708
PC(A)
CC(A),IR(A)
AR(A)
FI,AF,AT

TEXAS STATE TECHNICAL INSTITUTE
ACFT Pilot Training Program
James Connally Airport
Waco, TX 76705
PC(A)
CC(A),IR(A)
FI

TIGER AVIATION SERVICES, INC.
Crockett, David N.
Kleberg County Airport
P.O. Box 53
Kingsville, TX 78363
PC(A),PT(A)

VICTORIA AVIATION SERVICES, INC.
Rt. 6, Building 829
Victoria, TX 77901
PC(A)
CC(A),IR(A)
AR(A)
FI,AF

WES TEX AIRCRAFT, INC.
Rt. 3, Box 48
Lubbock, TX 79401
PC(A),PT(A)
CC(A),CT(A),IR(A)
PG(P)

WESTWIND AVIATION, INC.
1003 McKeever
Rosharon, TX 77583
PC(A)

YOUNG FLYING SERVICE
French Bowen Inc.
Valley Intl. Airport
Harlingen, TX 78550
PC(A)
CC(A),CT(A),IR(A)
AR(A)
FI,AF

UTAH

ALPINE AVIATION, INC.
P.O. Box 691
Provo Airport
Provo, UT 84601
PC(A)
CC(A),IR(A)
AR(A)
FI,AF

CENTRAL UTAH AVIATION
P.O. Box C
Provo, UT 84603
PC(A)

COLOR CANYONS AVIATION, INC.
P.O. Box 458
Cedar City, UT 84720
PC(A),PT(A)
CC(A),CT(A),IR(A)
AR(A),PG(P)
FI,AF,AT

INTERWEST AVIATION CORP.
AMF Box 22063
Salt Lake City, UT 84122
PC(A)
CC(A),CT(A),IR(A)
AR(A),PG(P)(I)(C)(F)
FI,AF,AT,PR

OGDEN AIR SERVICE
Wasatch Western Enterprises, Inc.
3911 S. Airport Rd.
Ogden, UT 84405
PC(A)
CC(A),IR(A)
FI,AF

SUNWEST AVIATION
Lindquist Investment Co., Inc.
3909 S. Airport Rd.
Ogden, UT 84405
PC(A)
CC(A),IR(A)
AR(A)
FI,AF,AT

THOMPSON BEECHCRAFT OF SALT LAKE CITY
369 N. 2370 West
Salt Lake City, UT 84116
PC(A)
CC(A),IR(A)

WASATCH HELICOPTER TRAINING CENTER
P.O. Box 435
Bountiful, UT 84010
PC(R)
CC(R),CT(R)
XL

WESTMINSTER COLLEGE
1840 S. 13th East
Salt Lake City, UT 84601
PG(P)(I)(C)(F)

VERMONT

MONTAIR FLIGHT SERVICE, INC.
1160 Airport Dr.
South Burlington, VT 05401
PC(A)
CC(A),IR(A)
AR(A)(L)
FI,AF,AT

VIRGINIA

AIR SHANNON, INC.
Shannon, Sidney L., Jr.
Shannon Airport
P.O. Box 509
Fredericksburg, VA 22401
PC(A),PT(A)
CC(A),IR(A)
AR(A)
FI,AF

CENTRAL VIRGINIA COMMUNITY COLLEGE
Wards Hill Station
Wards Rd. South, P.O. Box 4098
Lynchburg, VA 24502
PG(I)

COLGAN AIRWAYS
P.O. Box 1650
Manassas Municipal Airport
Manassas, VA 22110
PC(A)
CC(A),IR(A)
AR(A),PG(P)(I)(C)(F)(T)(R)
FI,AF,AT

DABNEY S. LANCASTER COMMUNITY COLLEGE
Rt. 60 West
Clifton Forge, VA 24422
PG(P)

DULLES AVIATION, INC.
P.O. Box 2169
Manassas Municipal Airport
Manassas, VA 22110
PC(A),PT(A)
CC(A),CT(A),IR(A)
AR(A),PG(P)(I)(C)
FI,AF

EXECUTIVE AIR
Bedford Flying Service, Inc.
Woodrum Field
Roanoke, VA 24012
PC(A)
CC(A)

EXECUTIVE AVIATION, INC.
Municipal Airport
Danville, VA 24541
PC(A)

FALLWELL AVIATION, INC.
Fallwell, W. C.
P.O. Box 937
Lynchburg, VA 24505
PC(A)
CC(A),IR(A)
FI

FLIGHT INTERNATIONAL, INC.
Patrick Henry Airport
Newport News, VA 23602
PC(A),PT(A)
CC(A),IR(A)
AR(A),PG(P)(I)(C)(F)(T)
FI,AF,AI,AT

GENERAL AVIATION, INC.
Rembold, E. Paul
Municipal Airport
P.O. Box 457
Danville, VA 24541
PC(A),PT(A)
CC(A),CT(A),IR(A)
AR(A)
FI,AF,AI,AT

HANOVER AVIATION CO.
604 Air Park Rd.
Ashland Hanover County Airport
Ashland, VA 23005
PC(A)
CC(A),IR(A)
AR(A),PG(P)(I)(C)(F)(R)(A)
FI,AF,AI

JANELLE AVIATION, INC.
Leesburg Municipal Airport
Rt. 1, P.O. Box 432
Leesburg, VA 22075
PC(A)

MID EASTERN AIRWAYS, INC.
P.O. Box 15400
Chesapeake, VA 23320
PC(A)
CC(A),IR(A)
AR(A)
FI,AF,AI,AT

NORFOLK NAVY FLYING CLUB
Norfolk Intl. Airport
1732 Dennys Rd.
Norfolk, VA 23502
AT

NORTHERN VIRGINIA COMMUNITY COLLEGE
Manassas Campus
6901 Sudley Rd.
Manassas, VA 22110
PG(P)

TIDEWATER COMMUNITY COLLEGE
District Office
Star Rt. 135
Portsmouth, VA 23703
PG(P)(I)(C)

VIRGINIA AVIATION
Aviation Resources Inc.
P.O. Box 4209
Lynchburg, VA 24502
PC(A)
CC(A),IR(A)
AR(A)
FI,AF,AI,AT,PR

WARING AVIATION, INC.
P.O. Box 7925
Charlottesville-Albemarle Airport
Charlottesville, VA 22906
PC(A)
CC(A),CT(A),IR(A)
AR(A)
FI,AF,AT

WOODBRIDGE AIRPORT, INC.
3314 Old Bridge Rd.
Woodbridge, VA 22192
PC(A)
CC(A),IR(A)
AR(A),PG(P)(I)(C)(F)(T)(R)
FI,AF,AT

WASHINGTON

AIRCRAFT SPECIALTIES, INC.
101 E. Reserve
Vancouver, WA 98661
PC(A)(R)
CC(A)(R),IR(A)
AR(A)(R)
FI

AUBURN FLIGHT SCHOOL
Renton Flight Service
P.O. Box 33061
Auburn, WA 98002
PC(A)
CC(A)
AR(A),PG(P)(I)(C)(F)(T)(R)

AVIATION TRAINING AND RESEARCH, INC.
ASA Ground School
7201 Perimeter Rd. South
Seattle, WA 98108
PC(A)
CC(A)
AR(A),PG(P)(I)(C)(F)(T)(R)
FI,AF,AI,AT

BAHR AERO
Bahr, Fredric
P.O. Box 48
Issaquah, WA 98027
PG(C)

BIG BEND COMMUNITY COLLEGE
Andrews & 24th Sts.
Moses Lake, WA 98837
PC(A)
CC(A),IR(A)
PG(P)(I)(C)(F)

CENTRAL WASHINGTON UNIVERSITY
Technology Dept.
Ellensburg, WA 98926
PG(P)(I)(C)

CLIFF HOWARD'S AVIATION
Howard, Cliff E.
8075 Perimeter Rd. South
Seattle, WA 98108
PC(A),PT(A)
CC(A),CT(A),IR(A)
AR(A),PG(P)(F)(T)(R)
FI,AF,AT

CLOVER PARK VO. TECH. INSTITUTE
4500 Steilacoom Blvd. SW
Tacoma, WA 98499
PC(A)
CC(A),IR(A)
AR(A),PG(P)(I)(C)(F)
FI,AF

FANCHER FLYWAYS, INC.
P.O. Box 412
Renton, WA 98055
PC(A)
CC(A),IR(A)
AR(A),PG(P)(I)(C)(F)
FI,AF,AT

FELTS FIELD AVIATION
P.O. Box 11877
Spokane, WA 99211
PC(A)

FLIGHTCRAFT, INC.
8285 Perimeter Rd. South
Seattle, WA 98108
PC(A),PT(A)
CC(A),CT(A),IR(A)
AR(A),PG(P)(I)(C)(F)
FI,AF,AT

GREEN RIVER COMMUNITY COLLEGE
12401 SE 320th St.
Auburn, WA 98002
PG(P)(I)(C)

INLAND AVIATION, INC.
P.O. Box 424
Ephrata, WA 98823
PC(A),PT(A)
CC(A),CT(A),IR(A)
AR(A)(R),PG(P)
AT

KENMORE AIR HARBOR, INC.
P.O. Box 64
Kenmore, WA 98028
PC(A),PT(A)
CC(A),IR(A)
AR(A),PG(P)(I)(C)(F)
FI,AF

MIDSTATE AVIATION, INC.
1101 Bowers Rd.
Ellensburg, WA 98926
PC(A)
CC(A),IR(A)
FI,AF

SNOHOMISH FLYING SERVICE, INC.
9807 Airport Way
Harvey Field
Snohomish, WA 98290
PC(A),PT(A)
CC(A),CT(A),IR(A)
AR(A)(R),PG(P)(I)(C)(F)
FI,AF,AT

SPANAFLIGHT
203 188th St. East
Spanaway, WA 98387
PC(A)
CC(A),IR(A)
AR(A),PG(P)(I)(C)(F)(T)
FI,AF

SPOKANE AIRWAYS, INC.
P. O. Box 19125
Spokane, WA 99219
PC(A),PT(A)
CC(A),CT(A),IR(A)
AR(A),PG(P)(I)(C)(F)

YANKEE COUNTRY FLIGHT CENTER, INC.
9115 NE 117th Ave.
Vancouver, WA 98662
PC(A)
CC(A),IR(A)
AR(A)
FI,AF

WEST VIRGINIA

KCI AVIATION
KCI Enterprises, Inc.
P.O. Box 849
Bridgeport, WV 26330
PC(A),PT(A)
CC(A),CT(A),IR(A)
AR(A)
FI,AF

LAWRENCE COUNTY AVIATION, INC.
Marshall University
Hal Green Blvd. and Third Ave.
Huntington, WV 25703
PG(P)

VALLEY AERO, INC.
P.O. Box 768
Martinsburg, WV 25401
PC(A)
CC(A),IR(A)
PG(P)(I)(C)
FI,AF

WISCONSIN

AERODYNE, INC.
4800 S. Howell Ave.
Milwaukee, WI 53207
PC(A)
CC(A),CT(A),IR(A)
AR(A)
FI,AF,AT

CARTER AIRCRAFT, INC.
Rt. 3
Pulaski, WI 54162
PC(A)
CC(A),IR(A)
FI,AF

CENTRAL WISCONSIN AVIATION, INC.
Central Wisconsin Airport
Mosinee, WI 54455
PC(A)
CC(A),IR(A)
FI,AF

CHAPLIN AVIATION, INC.
Rt. 1, County Trunk O
Sheboygan Falls, WI 53085
PC(A)
CC(A),IR(A)
AR(A)
FI,AF

FOND DU LAC SKYPORT, INC.
Rt. 5
Fond Du Lac, WI 54935
PC(A)
CC(A),IR(A)
AR(A)
FI,AF

FOUR LAKES AVIATION CORP.
3606 N. Stoughton Rd.
Madison, WI 53704
PC(A)
CC(A),IR(A)
AR(A)
FI,AF,AT

G A K AVIATION, INC.
Mitchell Aero, Inc.
923 E. Layton Ave.
Milwaukee, WI 53207
PC(A),PT(A)
CC(A),CT(A),IR(A)
AR(A)
FI,AF

GIBSON AVIATION SERVICE
3800 Starr Ave.
Eau Claire, WI 54701
PC(A)
CC(A),IR(A)
AR(A)
FI,AF

GRAN AIRE, INC.
9305 W. Appleton Ave.
Milwaukee, WI 53225
PC(A)
CC(A),IR(A)
AR(A)
FI,AF

HEILEMAN AIR SERVICES, LTD.
2709 Fanta Reed Rd.
La Crosse, WI 54601
PC(A)
CC(A),IR(A)
FI,AI

HODGE AERO, INC.
Rt. 7, Hwy. 51 South
Janesville, WI 53545
PC(A)
CC(A),IR(A)
AR(A)
FI,AF

MAXAIR INC.
Outagamie County Airport
Appleton, WI 54911
PC(A)
CC(A),IR(A)

MOREY AIRPLANE CO., INC.
Morey Airport
P.O. Box 8
Middleton, WI 53562
PC(A)
CC(A),IR(A)
AR(A)
FI,AF

SYLVANIA AIRPORT, INC.
2624 S. Sylvania Ave.
Sturtevant, WI 53177
CC(A),CT(A),IR(A)
AR(A)
FI,AF,AT

TRAGGIS AVIATION CORP.
P.O. Box 218
Municipal Airport
Hartford, WI 53027
PC(A)
CC(A),IR(A)
AR(A)
FI,AF

TWIN PORTS FLYING SERVICE, INC.
4804 Hammond Ave.
Superior, WI 54880
PC(A)
CC(A)(R),IR(A)
AR(A)

VIKING AVIATION, INC.
P.O. Box 877
Municipal Airport
La Crosse, WI 54601
PC(A)
CC(A),IR(A)
AR(A)
FI,AF,AT

WYOMING

CASPER AIR SERVICE, INC.
Natrona County International Airport
Casper, WY 82604
PC(A)(R)
CC(A),IR(A)
AR(A)
FI,AT

DISTRICT OF COLUMBIA

JET AMERICA INTERNATIONAL, INC.
General Aviation Terminal
Washington National Airport
Washington, DC 20001
AR(A)

PUERTO RICO

ISLA GRANDE FLYING SCHOOL AND SERVICE CORP.
P.O. Box C
Hato Rey, PR 00919
PC(A)
CC(A)

NATIONAL AVIATION ACADEMY, INC.
Puerto Rico International Airport
San Juan, PR 00913
PC(A)
PG(I)(C)

AVIATION MAINTENANCE TECHNICIAN SCHOOLS

These schools prepare you for your Airframe and/or Powerplant Licenses. The code key designating what each offers is at the beginning of the table.

Table 2-2. FAA-CERTIFICATED AVIATION MAINTENANCE TECHNICIAN SCHOOLS

CODES:

A	Airframe
P	Powerplant
A&P	Airframe and Powerplant

Note: Ratings are described in detail in FAR Part 147.

ALABAMA

ALABAMA AVIATION AND TECHNICAL COLLEGE
246 Club Manor Dr.
Mobile, AL 36615
A, P, A&P

ALABAMA AVIATION AND TECHNICAL COLLEGE
P.O. Box 1279
Ozark, AL 36361
A, P, A&P

NORTHWEST ALABAMA STATE TECHNICAL INSTITUTE
School of Aviation Technology
Marion County Airport
Hamilton, AL 35570
A, P, A&P

ALASKA

ANCHORAGE COMMUNITY COLLEGE
2811 Merrill Field Dr.
Anchorage, AK 99501
A, P, A&P

TANANA VALLEY COMMUNITY COLLEGE
Airframe and Powerplant Dept.
3750 Geist Rd.
Fairbanks, AK 99701
A, P, A&P

ARIZONA

CHOCHISE COLLEGE
Drawer L
Douglas, AZ 85607
A, P

ARKANSAS

PULASKI VO-TECH AVIATION MAINTENANCE SCHOOL
1600 W. Maryland Ave.
North Little Rock, AR 72116
A, P, A&P

SOUTHERN ARKANSAS UNIVERSITY
Technical Branch
Box 3048
East Camden, AR 71701
A, P, A&P

CALIFORNIA

CHAFFEY COLLEGE
5885 Haven Ave.
Alta Loma, CA 91701
A, P, A&P

CITY COLLEGE OF SAN FRANCISCO
Dept. of Aeronautics
San Francisco Intl. Airport
San Francisco, CA 94128
A, P

COLLEGE OF ALAMEDA
Box 2455, Airport Station
Oakland, CA 94614
A, P, A&P

COLLEGE OF SAN MATEO
1700 W. Hillsdale Blvd.
San Mateo, CA 94403
A, P

DEUEL VOCATIONAL INSTITUTION
P.O. Box 400
Tracy, CA 95376
A, P, A&P

GAVILAN COLLEGE
2310 San Felipe Rd.
Hollister, CA 95023
A, P

GLENDALE COMMUNITY COLLEGE DISTRICT
1500 North Verdugo
Glendale, CA 91208
P, A&P

JOHN A. O'CONNELL SCHOOL OF TECHNOLOGY
2905 21st Street
San Francisco, CA 94110
A, P

KINGS RIVER COMMUNITY COLLEGE
Reed and Manning Aves.
Reedley, CA 93654
A, P, A&P

LONG BEACH CITY COLLEGE
Business and Technology Div.
1305 E. Pacific Coast Hwy.
Long Beach, CA 90806
A, P, A&P

LOS ANGELES AIRPORT COLLEGE CENTER
9700 S. Sepulveda Blvd.
Los Angeles, CA 90045
A, P, A&P

MIRAMAR COLLEGE
San Diego Community College District
10440 Black Mountain Rd.
San Diego, CA 92128
A, P, A&P

MT. SAN ANTONIO COLLEGE
1100 N. Grand
Walnut, CA 91789
A, P, A&P

NORTH VALLEY OCCUPATIONAL CENTER
Aviation Trades Branch
16550 Saticoy St.
Van Nuys, CA 91406
A&P

NORTHROP INSTITUTE OF TECHNOLOGY
1155 W. Arbor Vitae
Inglewood, CA 90301
A, P, A&P

ORANGE COAST COLLEGE
2701 Fairview Rd.
Costa Mesa, CA 92626
A, P, A&P

PALO ALTO UNIFIED SCHOOL DISTRICT
50 Embarcadero Rd.
Palo Alto, CA 94301
A, P

SACRAMENTO CITY COLLEGE
3835 Freeport Blvd.
Sacramento, CA 95822
A, P, A&P

SAN BERNARDINO VALLEY COLLEGE
701 S. Mt. Vernon Ave.
San Bernardino, CA 92410
A, P, A&P

SAN JOSE STATE UNIVERSITY
125 S. Seventh St.
San Jose, CA 95192
A, P, A&P

SHASTA COLLEGE
1065 N. Old Oregon Trail
P.O. Box 6006
Redding, CA 96099
A, P, A&P

SIERRA ACADEMY OF AERONAUTICS
Technicians Institute
Oakland Intl. Airport
Oakland, CA 94614
A, P, A&P

SOLANO COMMUNITY COLLEGE SCHOOL OF AERONAUTICS
Box 246, Suisun Valley Rd.
Suisun City, CA 94585
A, P, A&P

COLORADO

COLORADO AERO TECH, INC.
10851 West 120th St.
Broomfield, CO 80020
A, P, A&P

COLORADO NORTHWESTERN COMMUNITY COLLEGE
General Delivery
Rangely, CO 81648
A, P

EMILY GRIFFITH OPPORTUNITY SCHOOL
Aircraft Training Center
8301 Montview Blvd.
Denver, CO 80220
A, P, A&P

CONNECTICUT

H. H. ELLIS REGIONAL VO-TECH SCHOOL
P.O. Box 149, Maple St.
Danielson, CT 06239
A, P, A&P

FLORIDA

BURNSIDE OTT AVIATION MAINTENANCE SCHOOL
14100 SW 129th St.
Tamiami Airport
Miami, FL 33186
A, P, A&P

BURNSIDE OTT AVIATION TECHNICAL SCHOOL
Pensacola Municipal Airport
P.O. Box 10735
Pensacola, FL 32504
A, P, A&P

EMBRY-RIDDLE AERONAUTICAL UNIVERSITY
Regional Airport
Daytona Beach, FL 32014
A, P, A&P

GEORGE T. BAKER AVIATION SCHOOL
3275 NW 42 Ave.
Miami, FL 33142
A, P, A&P

LEWIS M. LIVELY AREA VO-TECH AVIATION SCHOOL
500 N. Appleyard
Tallahassee, FL 32304
A, P, A&P

NATIONAL AVIATION ACADEMY, INC.
St. Petersburg Clearwater Airport
Clearwater, FL 33520
A&P

POMPANO ACADEMY OF AERONAUTICS
1101 NE 10th St.
Pompano Beach, FL 33060
A, P

RICE AVIATION
A and J Enterprises, Inc.
7991 Pembroke Rd.
Pembroke Pines, FL 33024
A, P

GEORGIA

ATLANTA AREA TECHNICAL SCHOOL
1560 Stewart Ave. SW
Atlanta, GA 30310
A, P, A&P

SOUTH GEORGIA TECHNICAL VOCATIONAL SCHOOL
P.O. Box 1088
Americus, GA 31709
A, P, A&P

HAWAII

HONOLULU COMMUNITY COLLEGE
AVMAT
402 Aokea Place
Honolulu, HI 96819
A, P, A&P

IDAHO

AERO TECHNICIANS, INC.
P.O. Box 7
Rexburg, ID 83440
A&P

IDAHO STATE UNIVERSITY
Municipal Airport, Hangar No. 4
Pocatello, ID 83201
A, P

ILLINOIS

BELLEVILLE AREA COLLEGE
Aviation Maint. Tech. Program
4950 Maryville Rd.
Granite City, IL 62040
A, P, A&P

BOARD OF EDUCATION CITY OF CHICAGO
Vo-Tech Dept. 6WS
1819 Pershing Rd.
Chicago, IL 60609
A, P, A&P

DUPAGE AREA VOCATIONAL EDUCATION AUTHORITY
301 N. Swift Rd.
Addison, IL 60101
P

LEWIS UNIVERSITY
Route 53
Romeoville, IL 60441
A, P, A&P

PARKS COLLEGE OF ST. LOUIS UNIVERSITY
Falling Springs Rd.
Cahokia, IL 62206
P, A&P

ROCK VALLEY COLLEGE AVIATION TECH
Greater Rockford Airport
6349 Falcon Rd.
Rockford, IL 61109
A, P, A&P

SOUTHERN ILLINOIS UNIVERSITY AVIATION TECH
Southern Illinois Airport
Carbondale, IL 62901
A, P, A&P

UNIVERSITY OF ILLINOIS INSTITUTE OF AVIATION
Univ. of Illinois Willard Airport
Savoy, IL 61874
A, P, A&P

VINCENNES UNIVERSITY
Aviation Maintenance Technology
Rt. 2, Box 40
Lawrenceville, IL 62439
A, P, A&P

INDIANA

PURDUE UNIVERSITY
DEPT. OF AVIATION TECHNOLOGY
Purdue University Airport
West Lafayette, IL 47906
A, P, A&P

IOWA

DES MOINES CENTRAL CAMPUS
1800 Grand
Des Moines, IA 50307
A&P

HAWKEYE INSTITUTE OF TECHNOLOGY
Merged Area Education VII
P.O. Box 8015
Waterloo, IA 50704
A, P, A&P

INDIAN HILLS COMMUNITY COLLEGE
Iowa Tech. Campus
Ottumwa Industrial Airport
Ottumwa, IA 52501
A, P, A&P

IOWA WESTERN COMMUNITY COLLEGE
2700 College Rd.
Council Bluffs, IA 51501
A, P, A&P

KANSAS

KANSAS TECHNICAL INSTITUTE
2409 Scanlan Ave.
Salina Municipal Airport
Salina, KS 67401
A, P

WICHITA AREA VO-TECH SCHOOL
Dagenais, Donald J.
Aviation Education Center
2021 S. Eisenhower
Wichita, KS 67209
A, P

KENTUCKY

SOMERSET STATE VO-TECH SCHOOL
RR 2, Box 512A
Somerset, KY 42501
A, P, A&P

LOUISIANA

DELGADO JUNIOR COLLEGE
State of Louisiana
615 City Park Ave.
New Orleans, LA 70119
A, P, A&P

SOWELA TECHNICAL INSTITUTE
3820 Legion St.
Lake Charles, LA 70601
A&P

MAINE

AIR TECH, INC.
16 Central Ave.
Limerick, ME 04048
A, P

MARYLAND

FREDERICK COMMUNITY COLLEGE
Route 3
Frederick, MD 21701
A, P, A&P

MASSACHUSETTS

EAST COAST AERO TECHNICAL SCHOOL, INC.
P.O. Box 412
Lexington, MA 02173
A, P, A&P

WENTWORTH INSTITUTE
550 Huntington Ave.
Boston, MA 02115
A, P, A&P

MICHIGAN

ANDREWS UNIVERSITY AIR PARK
Andrews Airport
Berrien Springs, MI 49104
A, P, A&P

BENJAMIN O. DAVIS, JR. AEROSPACE TECHNICAL H.S.
10200 Erwin Ave.
Detroit, MI 48234
A, P

DETROIT INSTITUTE OF AERONAUTICS
Willow Run Airport
Ypsilanti, MI 48197
A&P

GRAND RAPIDS SCHOOL OF THE BIBLE AND MUSIC
1331 Franklin SE
Grand Rapids, MI 49507
A&P

KIRTLAND COMMUNITY COLLEGE
Rt. 4, Box 59A
Roscommon, MI 48653
A, P, A&P

LANSING COMMUNITY COLLEGE
Capital City Airport, W. Hangar Dr.
Lansing, MI 48906
A, P

SOUTHWESTERN MICHIGAN COLLEGE
Dean, School of Technology
Cherry Grove Rd.
Dowagiac, MI 49047
A, P, A&P

WESTERN MICHIGAN UNIVERSITY
Transportation Technology Dept.
County Airport, Aviation Building
Kalamazoo, MI 49001
A, P

MINNESOTA

MINNEAPOLIS AREA VO-TECH INSTITUTE
Aviation Center
10100 Flying Cloud Dr.
Eden Prairie, MN 55344
A, P, A&P

THIEF RIVER FALLS AREA TECHNICAL SCHOOL
General Delivery
Thief River Falls, MN 56701
A, P, A&P

WINONA AREA TECHNICAL SCHOOL
1250 Homer Rd.
Winona, MN 55987
A&P

MISSISSIPPI

GOLDEN TRIANGLE VO-TECH CENTER
Aviation Facility
P.O. Box 789
Columbus, MS 39701
A, P, A&P

HINDS JUNIOR COLLEGE
General Delivery
Raymond, MS 39154
A, P, A&P

NORTHWEST MISSISSIPPI JUNIOR COLLEGE
8750 Deerfield Dr.
Olive Branch, MS 38654
A&P

MISSOURI

AERO MECHANIC SCHOOL
838 Richard Rd.
Kansas City, MO 64116
A, P, A&P

LINN TECHNICAL COLLEGE
General Delivery
Linn, MO 65051
A, P, A&P

MAPLE WOOD COMMUNITY COLLEGE
2601 NE Barry Rd.
Kansas City North, MO 64156
A, P

O'FALLON TECHNICAL CENTER
5101 McRee
St. Louis, MO 63110
A, P, A&P

School of the Ozarks
Aviation Maint. Tech. School
Point Lookout, MO 65726
A, P

MONTANA

Helena Vo-Tech Center
Aero Division
2300 E. Poplar
Helena, MT 59601
A, P, A&P

NEBRASKA

Western Nebraska Technical College
General Delivery
Sidney, NE 69162
A, P

NEVADA

Southern Nevada Vo-Tech Center
5710 Mountain Vista Dr.
Las Vegas, NV 89120
A, P, A&P

NEW JERSEY

Cumberland County Vo-Tech Center
RR 8, Box 45, Bridgeton Ave.
Bridgeton, NJ 08302
P

Teterboro School of Aeronautics, Inc.
80 Moonachie Ave.
Teterboro Airport
Teterboro, NJ 07608
A, P, A&P

NEW MEXICO

Eastern New Mexico University
Roswell Campus
P.O. Box 6000
Roswell, NM 88201
A&P

NEW YORK

Academy of Aeronautics
LaGuardia Airport
Flushing, NY 11371
A, P, A&P

Aviation High School
36th St. and Queens Blvd.
Long Island City, NY 11101
A, P, A&P

B O C E S II
Islip MacArthur Aviation Center
2965 Smithtown Blvd.
Ronkonkoma, NY 11779
A

Burgard Vocational High School
400 Kensington Ave.
Buffalo, NY 14214
A, P, A&P

East NY Vo-Tech High School
1 Wells St.
Brooklyn, NY 11208
P

Lewis Wilson Technological Center
Grand Blvd. and Lucon Dr.
Deer Park, NY 11729
A, A&P

Mid Westchester Center for Occupational Education
65 Grasslands Rd.
Valhalla, NY 10595
P

Nassau Technological Center
610 Hicksville Rd.
Bethpage, NY 11714
A, P, A&P

Riverside School of Aeronautics, Inc
Riverside Airport, P.O. Box 444
Utica, NY 13503
A&P

NORTH CAROLINA

GUILFORD TECHNICAL COMMUNITY COLLEGE
501 W. Washington St.
Greensboro, NC 27401
A, P, A&P

MISSIONARY AVIATION INSTITUTE, INC.
Piedmont Bible College
Sugar Valley Airport
Rt. 2, Box 381
Mocksville, NC 27028
A

WAYNE COMMUNITY COLLEGE
Caller Box 8002
Goldsboro, NC 27530
A&P

NORTH DAKOTA

DAKOTA AERO TECH, INC.
P.O. Box 5534, State University Sta.
Fargo, ND 58102
A, P, A&P

OHIO

AVIATION HIGH SCHOOL
Cleveland Public Schools
4101 N. Marginal Rd.
Cleveland, OH 44114
A, P, A&P

CINCINNATI TECHNICAL COLLEGE
Aviation Dept.
3520 Central Pkwy.
Cincinnati, OH 45223
A, P, A&P

COLUMBUS TECHNICAL INSTITUTE
5355 Alkire Rd.
Columbus, OH 43228
A, P, A&P

MACOMBER VOCATIONAL HIGH SCHOOL
Aerospace Center
1501 Monroe St.
Toledo, OH 43624
A

MONTGOMERY CO. JOINT VOCATIONAL HIGH SCHOOL
6800 Hoke Rd.
Clayton, OH 45315
A, P, A&P

W. C. SMITH VOCATIONAL HIGH SCHOOL
7300 N. Palmyra Rd.
Canfield, OH 44406
A&P

OKLAHOMA

AERONAUTICAL TECHNOLOGY
Technical Institute
Oklahoma State Univ.
Stillwater, OK 74074
A, P, A&P

FOSTER ESTES AREA VO-TECH CENTER
4901 S. Bryant
Oklahoma City, OK 73109
P

GORDON COOPER AREA VO-TECH SCHOOL
P.O. Drawer 848
Shawnee, OK 74801
A

INDIAN CAPITAL AREA VO-TECH SCHOOL
Rt. 6, Box 206
Muskogee, OK 74401
A

J. T. AUTRY AREA VO-TECH CENTER
1201 W. Willow
Enid, OK 73701
A, P

SPARTAN SCHOOL OF AERONAUTICS, INC.
8820 E. Pine St.
Tulsa, OK 74151
A, P, A&P

TULSA CO. AREA VO-TECH SCHOOL
1200 W. 36 St. North
Tulsa, OK 74127
A

OREGON

LANE COMMUNITY COLLEGE
4000 E. 30th Ave.
Eugene, OR 97405
A, P

PORTLAND COMMUNITY COLLEGE
17705 NW Springville Rd.
Portland, OR 97229
A&P

PENNSYLVANIA

UPPER BUCKS CO. TECHNICAL SCHOOL
Box 392, Old Milford Sq. Rd.
Quakertown, PA 18951
A&P

PITTSBURGH INSTITUTE OF AERONAUTICS
Allegheny County Airport
West Mifflin, PA 15122
A, P, A&P

QUAKER CITY INSTITUTE OF AVIATION, INC.
2563-69 Grays Ferry Ave.
Philadelphia, PA 19146
A, P, A&P

WILLIAMSPORT AREA COMMUNITY COLLEGE
1005 W. Third St.
Williamsport, PA 17701
A, P, A&P

SOUTH CAROLINA

BOB JONES UNIVERSITY
1700 Wade Hampton Blvd.
Greenville, SC 29614
A&P

FLORENCE DARLINGTON TECHNICAL COLLEGE
P.O. Drawer 8000
Florence, SC 29501
A, P

NORTH AMERICAN INSTITUTE OF AVIATION OF SOUTH CAROLINA
Conway Horry County Airport
P.O. Box 680
Conway, SC 29526
A&P

TRIDENT TECHNICAL COLLEGE
7000 Rivers Ave.
Charleston, SC 29411
A, P

SOUTH DAKOTA

LAKE AREA VO-TECH INSTITUTE
Municipal Airport
Watertown, SD 57201
A, P

TENNESSEE

MEMPHIS AREA VO-TECH SCHOOL
Aviation Division
2752 Winchester Rd.
Memphis, TN 38116
A, P, A&P

MOODY AVIATION
Moody Bible Institute of Chicago
Box 429, Municipal Airport
Elizabethton, TN 37643
A, P, A&P

TEXAS

CENTRAL TEXAS COLLEGE
Hwy. 190 West
Killeen, TX 76541
P

HALLMARK INSTITUTE OF TECHNOLOGY
Hallmark Aero Tech., Inc.
1130 99th St.
San Antonio, TX 78214
A, P, A&P

LETOURNEAU COLLEGE
Dept. of Aeronautical Technology
P.O. Box 7001
Longview, TX 75601
A&P

LOVE A AND P TRAINING CENTER
Ruffin, Pauline
105 Dividend Ct.
Arlington, TX 76102
A, P, A&P

RICE AVIATION
A and J Enterprises, Inc.
205 Brisbane
Houston, TX 77061
A&P

SKYLINE HIGH SCHOOL
Career Development Center
7777 Forney Rd.
Dallas, TX 75227
A, P

T S T I AMARILLO AVIATION
Maintenance Technician
P.O. Box 11035
Amarillo, TX 79111
A&P

TARRANT CO. JR. COLLEGE
4801 Marine Creek Parkway
Ft. Worth, TX 76179
A, P, A&P

TEXAS AERO TECH.
7326 Airfreight Lane
Dallas, TX 75235
A, P, A&P

TEXAS STATE TECHNICAL INSTITUTE
James Connally Campus
Waco, TX 76705
A, P

UTAH

DIXIE COLLEGE
255 South 700 East
St. George, UT 84770
A&P

RAINBOW AVIATION, INC.
P.O. Box 3485
Logan, UT 84321
A&P

UTAH STATE, UNIVERSITY OF AGRICULTURE AND APPLIED SCIENCES
College Hill
Logan, UT 84321
A, P, A&P

UTAH TECHNICAL COLLEGE AT SALT LAKE
4600 S. Redwood Rd.
Salt Lake City, UT 84131
A, P

VIRGINIA

RICE AVIATION OF VIRGINIA
A and J Enterprises, Inc.
911 Live Oak Dr.
Suites 103 and 104
Chesapeake, VA 23320
A, P

WASHINGTON

BIG BEND COMMUNITY COLLEGE
Building 4103, North Campus
Moses Lake, WA 98837
A, P, A&P

CLOVER PARK VO. TECH. INSTITUTE
4500 Steilacoom Blvd. SW
Tacoma, WA 98499
A, P, A&P

EVERETT COMMUNITY COLLEGE
2915 112th SW, Building 512
Everett, WA 98204
A, P, A&P

SO. SEATTLE COMMUNITY COLLEGE
6000 16th Ave. SW
Seattle, WA 98106
A, P, A&P

SPOKANE COMMUNITY COLLEGE
North 1810 Greene
Spokane, WA 99207
A, P, A&P

WISCONSIN

BLACKHAWK TECHNICAL INSTITUTE
2228 Center Ave.
Janesville, WI 53545
A, P, A&P

GATEWAY TECHNICAL INSTITUTE
3520-30th Ave.
Kenosha, WI 53140
A, A&P

MILWAUKEE AREA TECHNICAL COLLEGE
1015 N. 6th St.
Milwaukee, WI 53203
A, P, A&P

WYOMING

CHEYENNE AERO TECH
3801 Morrie Ave.
Cheyenne, WY 82001
A, P, A&P

DISTRICT OF COLUMBIA

UNIVERSITY OF THE DISTRICT OF COLUMBIA
Washington National Airport Campus
Washington, DC 20001
A, P, A&P

PUERTO RICO

MIGUEL SUCH METROPOLITAN VOCATIONAL SCHOOL
P.O. Box 21837, University Station
Rio Piedras, PR 00931
A, P

AIRCRAFT DISPATCHER AND FLIGHT ENGINEER

Table 2-3 lists the schools, by region, that offer approved courses in aircraft dispatching and/or being a flight engineer for the four categories of basic, reciprocating, turboprop, and turbojet flight engineer. Again, the codes key is at the top of the table.

Table 2-3. SCHOOLS WITH FAA-APPROVED COURSES FOR AIRCRAFT DISPATCHER AND FLIGHT ENGINEER

COURSE CODES:

AD	Aircraft Dispatcher
FEB	Flight Engineer—Basic
FER	Flight Engineer—Reciprocating
FEP	Flight Engineer—Turboprop
FEJ	Flight Engineer—Turbojet

REGION	NAME AND ADDRESS	COURSE
Alaskan	None	None
Central	Trans World Airlines, Inc. Training Center 1307 Baltimore Kansas City, MO 64105	FEB/FEJ (Ground/Flt)

REGION	NAME AND ADDRESS	COURSE
Eastern	Academics of Flight 43-34 43rd St. Sunnyside, NY 11104	AD FEB/FEJ (Ground/Flt)
	Airline Operations Training 36 S. Station Plaza Great Neck, NY 11021	AD FEB/FEJ
European	None	None
Great Lakes	International Airlines Academy, Inc. d/b/a Transcontinental Airlines, Inc. Willow Run Airport Ypsilanti, MI 48197	FEB/FER/ FEP/FEJ (Ground/Flt)
	*Northwest Airlines, Inc. Minneapolis-St. Paul International Airport St. Paul, MN 55111	AD FEB/FEJ (Ground/Flt)
	Purdue University Dept. of Aviation Technology Purdue Airport West Lafayette, IN 47907	FEB/FEJ (Ground/Flt)
	*Republic Airlines, Inc. 7500 Airline Drive Minneapolis, MN 55450	AD FEB/FEJ (Ground/Flt)
	*Zantop International Airlines, Inc. Willow Run Airport Ypsilanti, MI 48197	FER/FEP (Ground/Flt)
New England	None	None
Northwest Mountain	Central Wash. State College Ellensburg, WA 98926	FEB/FEJ (Ground only)
	Metropolitan State College 250 W. 14th Denver, CO 80204	FEB/FEJ (Ground only)
	United Air Lines, Inc. Flight Training Center Stapleton Intl. Airport Denver, CO 80207	FEB/FEJ (Ground/Flt)

REGION	NAME AND ADDRESS	COURSE
Southern	Burnside-Ott Aviation Training Center Opa Locka Municipal Airport Opa Locka, FL 33054	FEB/FEJ (Ground/Flt)
	Flight International, Inc. Peachtree-Dekalb Airport 1954 Airport Road Suite 283 Atlanta, GA 30341	FEB/FEJ (Ground/Flt)
	Miami-Dade Jr. College (North Campus) 11380 NW 27th Ave. Miami, FL 33167	FEB/FEJ (Ground only)
	Sheffield Schools 5559 NW 36th St. Miami Springs, FL 33166	AD FEB/FEJ (Ground/Flt)
Southwest	American Airlines, Inc. Flight Training Academy Fort Worth, TX 76125	FEB/FEJ (Ground only)
	**American Flyers, Inc. Airpark Branch P.O. Box 3241 Ardmore, OK 73401	FEB/FEJ (Ground/Flt)
	Braniff Education Systems, Inc. P.O. Box 35001 Dallas, TX 75235	FEB/FEJ (Ground/Flt)
	Mountain View College 4849 W. Illinois Dallas, TX 75211	AD
	**Southeastern State College Dept. of Professional Aviation Durant, OK 74701	FEB/FEJ (Ground/Flt)
	United States Air Force 443rd Technical Squadron Altus AFB, OK 73521	FEB/FEJ (Ground/Flt)
Western/ Pacific	Accelerated Ground Training, Inc. 3200 Airport Avenue Santa Monica, CA 90405	FEJ (Flight only)

REGION	NAME AND ADDRESS	COURSE
Western/ Pacific	Airline Training Institute 2121 S. El Camino Real Suite 502 San Mateo, CA 94403	FEB/FEJ
	California Airlines Institute Aircraft Dispatcher School 2409 Sepulveda Blvd. Manhattan Beach, CA 90266	AD
	Flight Training Devices 312 East Imperial Ave. El Segundo, CA 90245	FEJ (Flight only)
	Flying Tiger Line, Inc. 7401 World Way West Los Angeles Intl. Airport Los Angeles, CA 90009	FEB/FEJ (Ground/Flt)
	Fowler Aeronautical Service 3031 W. Burbank Blvd. Burbank, CA 91505	FEB/FEJ/FEP (Ground/Flt)
	Mt. San Antonio College 1100 Grand Ave. Walnut, CA 91789	FEB/FEJ (Ground only)
	Sierra Academy of Aeronautics Oakland Intl. Airport Oakland, CA 94614	FEB/FEJ
	Western Airlines, Inc. P.O. Box 92005 Los Angeles, CA 90009	FEB/FEJ (Ground/Flt)

*Not available to the public. Approved as a Part 121 training program only.

**These fully approved schools have contractual arrangements with American Airlines, Fort Worth, Texas, to conduct the required flight training.

FOUR-YEAR COLLEGES

According to the American Association of Airport Executives, the educational background to prepare for a career in airport management or a similar career would ideally include courses in engineering, business and personnel management, business law, and aviation legislation. A number of schools offer a curriculum in aviation management or aviation administration that fulfills the above criteria.

The following is a listing of four-year colleges that fall into this category. For a more detailed listing, refer to the "Collegiate Aviation Directory" by writing to the University Aviation Association, P.O. Box 2321, Auburn, AL 36830. (Many of these colleges are also listed in the FAA-certificated listings of schools.)

Table 2-4. Four-Year Colleges with Aviation Degrees

INSTITUTION	ADDRESS
Auburn University	Auburn, AL 36845
Troy State University	Troy, AL 36081
Arizona State University	Tempe, AZ 85281
California State University	Long Beach, CA 90840
San Jose State University	San Jose, CA 95114
University of California	Berkley, CA 94720
Metropolitan State College	Denver, CO 80204
University of New Haven	West Haven, CT 06516
Wilmington College	New Castle, DE 19720
Embry-Riddle Aeronautical University	Daytona Beach, FL 32014 or Prescott, AZ 86301
Florida International University	Miami, FL 33199
Florida Institute of Technology	Melbourne, FL 32901
Boise State University	Boise, ID 83725
Aerospace Institute	Chicago, IL 60610
Lewis University	Romeoville, IL 60441
Parks College of St. Louis	Cahokia, IL 62206
Indiana State University	Terra Haute, IN 47809
Purdue University	West Lafayette, IN 47907
University of Dubuque	Dubuque, IA 52001
Kansas Newman College	Wichita, KS 67213
Southwestern College	Winfield, KS 67156
Louisiana Technical University	Ruston, LA 71272
Northeast Louisiana University	Monroe, LA 71209
Western Michigan University	Kalamazoo, MI 49008
Mankato State University	Mankato, MN 56001
Central Missouri State University	Warrensburg, MO 64093
Nathaniel Hawthorne College	Antrim, NH 03440
Daniel Webster College	Nashua, NH 03060
University of Albuquerque	Albuquerque, NM 87140
Dowling College	Oakdale, NY 11769
St. Francis College	Brooklyn NY 11201
University of North Dakota	Grand Forks, ND 58202
Kent State University	Stow, OH 44224
Ohio State University	Columbus, OH 43210
Oklahoma City University	Oklahoma City, OK 73106
Oklahoma State University	Stillwater, OK 74074
Augusta College	Souix Falls, SD 57197
National College	Rapid City, SD 57709
Middle Tennessee State University	Murfreesboro, TN 37132
American Technological University	Killeen, TX 76541
University of Houston	Houston TX 77004
University of Texas—Permian Basin	Odessa, TX 79762
Salem College	Salem, WV 26426

AB INITIO FLIGHT TRAINING

Considering the number of experienced pilots that will be retiring within the next ten years, the need for a new source of airline pilots should be a surprise to no one. The military has long been a reliable source of airline pilots in the past. Other traditional sources include corporate pilots and flight instructors. Although these sources are still supplying pilots to the airlines, it is predicted that the demand for pilots will far outweigh the dwindling supply offered by these sources.

Given these dim statistics, some airlines have taken the initiative to find a workable solution to the situation before it becomes a major problem. Three major airlines have now taken it upon themselves to train pilots literally from the ground up, using established aviation colleges.

This new concept in pilot training is known as *Ab Initio* airline training. Literally translated, Ab Initio means "from the beginning" in Latin. This program differs from the aviation programs typically offered by aviation colleges such as Embry-Riddle Aeronautical University in that it is geared specifically toward the airline industry. Most modern aviation programs cover a broad range of topics in the aviation field. The Ab Initio program is designed to teach "zero-time," or "low-time" pilots everything that they need to know in roughly 300 flight hours (in most cases). After graduation, the individual is qualified to interview for a pilot position with prospective airlines. The program does not guarantee the graduating student a job, but the airlines will consider graduation from the program in lieu of the minimum hour requirements usually necessary for airline employment.

The precedent for this program already exists with the military services and many foreign airlines, including British Airways. Ab Initio training was actually tried in the United States by Trans World Airlines back in the mid 1960s. During this short-lived period when there was a pilot shortage in the U.S., Trans World Airlines hired zero-time university graduates, paid for all the required commercial pilot training, and put them to work as pilots and flight engineers. This program was reported to be quite successful, and many of those who were trained are still actively flying with the airlines. After the end of the Vietnam war, the supply of qualified pilots was replenished, and TWA abandoned the pilot development program and selected experienced aviators instead. The fact of the matter is that the Ab Initio concept is a valid form of pilot training that seems to be one of the most logical answers to the lack of qualified aviators. A major difference between the Ab Initio program and the aeronautical universities is the affiliation between the school and the airline. The three airlines mentioned earlier that are involved in the Ab Initio program to date are Northwest, Eastern, and United.

Northwest Aerospace Training Corporation (NATCO), the training arm of Northwest Airlines, has affiliated itself with the University of North Dakota's Center for Aerospace Studies (CAS). The university has had aerospace programs for many years; however, the new NATCO/CAS organization will be a separate entity with a course curriculum that is specifically airline oriented.

United Airlines is affiliated with Southern Illinois University at Carbondale, Illinois (SIUC). This program actually puts students into the airline's flight training facility at Denver, where they work as interns. The basic training leads to commercial, instrument, multi-engine, and flight instructor's ratings. Total flight time gained in this program is between 400 and 600 hours. To be accepted into the United/SIUC Ab Initio program, a student must apply from within SIUC's aviation management department, have a grade point average of 2.5, pass an FAA class I physical, and have earned an associate degree in Aviation Flight.

Eastern, like Northwest and United, has developed a relationship with teaching institutions, however Eastern is using two-year colleges rather than four-year colleges. At present, Eastern is involved in an Ab Initio program with three different colleges and is negotiating with various others. Present colleges include Miami Dade Community College in Florida, Aims Community College in Greeley, Colorado, and San Jacinto College in Pasadena, Texas. The graduating student will earn an Associate Degree in Aviation Science as well as a commercial, multi-engine, and instrument rating, about 250 hours of flight time and must pass a flight engineer written exam.

Although the Ab Initio concept of training seems like the answer to every hopeful pilot's dreams, the cost of the program is worth consideration. Getting a good education is not cheap no matter where the individual goes. For example, at Embry-Riddle Aeronautical University, the average four-year college degree in aviation costs about $45,000. This price includes a B.S. degree in Aeronautical Science (or equivalent) and about 250 flight hours, with all the appropriate flight ratings included. (Being a 1982 Embry-Riddle graduate, I feel the instruction to be worth the time and money at any price.) The typical Ab Initio program in a four-year college would cost in the $40,000 range, depending on the college attended. This cost includes an appropriate four-year degree (in some cases) and the flight ratings necessary for the entry-level airline pilot. The program that Eastern is sponsoring is a bit less expensive because two-year colleges are utilized in place of four-year colleges. The average cost of the two-year Ab Initio program is between $9,000 and $15,000. The graduating student will have an appropriate two-year degree and the flight ratings necessary for employment.

Perhaps the most ambitious Ab Initio program, and the one that most resembles the European programs, is the one being sponsored by Northwest Aerospace Training Corporation (NATCO) and the University of North Dakota. Lets take a closer look at its background and some of the finer details of the program.

In September of 1987, NATCO broke ground on a new 6-million-dollar training facility located adjacent to the current center for Aerospace Sciences building in Grand Forks, North Dakota. This 6-million-dollar center was funded jointly by NATCO and the FAA. The building houses computer interactive classrooms, computer-based instruction labs, briefing rooms, and facilities for a full range of simulators and cockpit procedures trainers. The new facility became available for occupancy in January of 1989. Training at this facility will be comprehensive, with 700 to 900 hours of academic instruction and approximately

300 hours of simulator and flight training. Students train in a variety of simulators and aircraft with advanced options available for turboprop, light turbojet, and transport category (DC-9, B-727) equipment. Airline operations and training concepts are emphasized in virtually all phases of training. The goal is to produce graduates of the highest possible caliber—pilots who are fully prepared for the airline cockpit.

The school evaluates all applications to ensure that candidates meet airline educational, physical, and psychological hiring standards. Initial application requirements include a four-year college degree and an FAA Class I airman's medical certificate. Candidates meeting initial screening prerequisites are invited to Minneapolis to undergo physical and psychological evaluations and an interview. Successful candidates are notified immediately following the screening and selection process. The full package pilot training program is estimated to be 18 months long.

To the south, just outside of Minneapolis, NATCO is busy building its primary training base next to the new world headquarters of Northwest Airlines. This 20-million-dollar project is independent of the NATCO/University of North Dakota venture. The center, when fully expanded to full capacity, will house 32 full-flight simulators. Nine new high-technology simulators were ordered in the largest single simulator procurement in the history of commercial aviation. The simulator contract, representing a 50-million-dollar investment, includes construction of simulators for two of Northwest's airliners of the future: the Boeing 747-400 and the Airbus A320-200. In addition, NATCO has contracted to modify and upgrade the 12 simulators it currently owns. This impressive facility will be used to train and upgrade the bulk of Northwest's crew of pilots. These facilities will also be used to train beginning pilots for other airlines, governments, and corporations on a contract basis.

Many aspects of the Ab Initio programs now being offered are still in the planning stages. For specific information on a particular program, contact the school or training institution in question.

3

Engineering and Technical Support

TA-4J Skyhawk

This is the dawning of a new era in the field of aerospace engineering. Man now travels around the globe at supersonic speeds, transports tons of passengers and cargo to hundreds of thousands of destinations each day, and with each passing day the thought of extended space travel becomes more and more realistic.

This type of achievement doesn't just happen. Behind the scenes are thousands of skilled engineers who devote their lives to the advancement of aerospace.

Perhaps more than ever before, the opportunity to be a part of this aerospace revolution is possible if you know where to look. As the field of aerospace becomes more and more complex, people have become more specialized in their field of expertise. It now seems that there is an engineer for every phase of the equipment design and integration. In many cases, choosing a position can be as complicated as choosing an employer.

This chapter looks at some of the career fields offered by aerospace employers in today's market and tries to define the job responsibilities associated with these positions. Desired training and education will also be addressed to help the individual prepare for such a position.

At the end of this chapter is an alphabetical listing of potential employers (Table 3-1). Names, addresses, phone numbers, and other helpful information is included for the job hunter's reference.

JOB DESCRIPTIONS

The following is an alphabetical listing of positions typically found in many aerospace organizations. All of the job descriptions in this chapter are courtesy of the Boeing Company.

Acoustics (Degree in AE, EE, ME, or Physics)

Design, calibrate, and apply high-power transducers in industrial processing. Study ultrasonic cleaning techniques and noncontacting measurements of surface velocities. Develop electronic applications to acoustic data measurements and analysis systems; develop and apply special microphone calibration procedures for acoustical excitation and vibration. Perform studies in the field of psychoacoustics, which includes the development of experimental techniques for determining subjective responses to broadband noise and developing changes in aircraft design to prevent, reduce, or remove noise characteristics causing crew fatigue and other problems. Design electromechanical devices and develop high-intensity noise-testing procedures and devices; evaluate the results of structural and electronic response to this testing and help develop new aircraft and missile designs, devices, or materials to prevent structural and other damage caused by this noise. Develop new or improved noise-reducing materials and devise applications of these materials to company products or noise problem areas. (Also see *Noise Technology*.)

Aerodynamic Configuration Design and Analysis (Degree in AE)

By using advanced theoretical techniques, wind tunnel tests, and flight tests, determine the external shape of an aircraft, missile, or space vehicle with the objective of developing a configuration that will provide the best possible performance in terms of range, altitude, payload, and speed, and also have excellent flying qualities. In pursuing this objective, cognizance must be taken of such factors as aerodynamic heating, aeroelasticity, stability and control, allowable CG locations, and many other specialized problems for each specific design. This involves analyzing and studying basic requirements affecting the configuration, assembling the best available knowledge and information on pertinent configurations or their components, and developing new analytical and empirical information to establish basic characteristics of the new configuration.

Aerodynamic Stability and Control (Degree in AE or experience in flight vehicle stability and control)

Analyze flight vehicles from the standpoint of aerodynamic stability and control criteria for piloted and unmanned vehicles through the use of wind tunnel tests, theoretical analysis, and flight simulation. Determine vehicle configuration and control concepts that will meet these criteria while causing a minimum penalty to vehicle performance. Establish the design requirements of the desired concepts, in close cooperation with system and structural designers. Estimate the flying qualities of proposed vehicles and establish aerodynamic derivatives to use in autopilot or guidance system design. During detail design and construction of the vehicle, see that all criteria for aerodynamic control and stability are met. Establish requirements to obtain stability and control data during flight tests and make any necessary adjustments or refinements in the system to achieve the desired flight characteristics. Analyze and document flight test results to verify compliance with vehicle flying qualities design requirements. Develop and validate flight simulator aerodynamic math model.

Aerodynamics—General (Degree in AE or experience in theoretical or applied aerodynamics)

Conduct performance analysis to determine general vehicle (airplane or missile) configurations to meet customer requirements. Perform analytical studies and wind tunnel tests to develop and refine complete configurations as well as components such as wings, bodies, fins, and engine nacelles. This development can include studies in the basic nature of such phenomena as shock waves, boundary layers, and potential flow fields. Utilize the results of these analyses and tests to predict exact performance of the vehicle to be offered to the customer. Follow the detail design and construction of the vehicle to see that aerodynamic criteria are met. Determine requirements for flight testing items of aerodynamic interest and analyze the results of these tests to establish the vehicle performance and to find areas of further aerodynamic improvement. Advise the customer through handbooks and direct contact on how to achieve optimum aerodynamic performance from the product sold to him.

Aeroelasticity (Degree in AE, CE, EE, EM or ME)

Solve elementary flutter and vibration problems and perform complex calculations and research in aeroelasticity, involving the following types of activities: Conduct wind tunnel tests of scaled elastic flutter models, evaluating data and making suitable recommendations to project or preliminary design personnel; conduct flight flutter tests of full-scale aircraft in order to verify that margins of safety with respect to flutter are adequate; measure environmental vibration on aircraft; evaluate data; make recommendations for vibration testing of equipment; and conduct experimental and analytical research to develop improved methods of flutter prediction. Evaluate the latest in steady and unsteady aerodynamics research and develop methods for applying these new developments to a variety of aeroelastic problems.

Aerothermodynamics (Degree in AE or ME)

Perform theoretical and experimental analyses of thermodynamics of combustion engine cycles applicable to aircraft, missile, and space vehicle programs involving ramjet, turbojet, rocket, and advanced propulsion concepts. Develop methods of analysis for predicting heat transfer to material surfaces that are exposed to high-speed gas flows (wings and other external surfaces exposed to high-temperature gaseous products of combustion). This involves such functions as defining radiative and convective heat transfer distributions of hypersonic vehicles including regions of shock wave boundary layer interaction regions. Calculate the temperature response of aircraft, missiles, and space vehicles subjected to these heating environments. Determine thermal protection requirements needed to maintain acceptable temperatures. Explore possible techniques for reducing aerodynamic heating and its deleterious side effects on boundary layer growth. Study boundary layer separation phenomenon and associated effects on pressure and heat transfer distributions. Perform trade studies to determine the effects of configuration geometry and flight condition on heat transfer, including consideration of pressure gradients, surface condition, and real gas effects.

Antennas and Radomes (Degree in EE or Physics)

Perform theoretical studies as well as laboratory development evaluations of actual hardware for aircraft, spacecraft, and advanced weapons systems. This includes gas, solid, and liquid dielectric research to help solve high-voltage antenna-breakdown problems that occur during passage through the ionosphere; design of high-temperature antennas, development of large-aperture, electrically scanned antennas for high-performance air-to-surface missiles, space vehicles, and early warning systems; development and test communications and navigation antennas for commercial aircraft; development of broadband and band-pass radomes for advanced electronic reconnaissance systems, electronic countermeasure antennas for advanced bomber aircraft and weather radar, and navigation system radomes for commercial aircraft.

Applications Research (Advanced degree in EE and experience in logic design, switching circuits or data transmission and recording techniques)

Perform research in data transmission, recording, translating, reduction, and presentation equipment and techniques required for flight test data reduction.

Applied Mathematics (Advanced degree in Computer Science, Mathematics, or Physics)

Formulate mathematical models or other descriptions of engineering or decision processes. Use appropriate mathematical techniques to analyze such models and obtain problem solutions through analysis and/or numerical methods. A wide range of mathematical techniques may be employed, such as classical analysis, probability and statistics, control theory, graph theory, optimization, stochastic processes, numerical analysis, queuing theory, or approximation theory. The problems may arise from areas such as acoustics, aerodynamics, electronics, electromagnetics, heat transfer, propulsion, structural analysis, operations research, logistics, navigation and guidance, communications optics, stress analysis, or computer performance analysis.

Artificial Intelligence (Degree in Computer Science, Math, EE or Physics; Graduate study in AI desirable)

Perform basic and advanced research to develop and advance AI technology. Apply advanced AI technology to both military and commercial products. Develop AI methodology and tools. Problems encompass most engineering disciplines and require an overall system approach to define optimum solutions. Individual assignments include research and application of one or more AI subsets, such as knowledge-based systems engineering, expert systems, intelligent robotics, vision and image processing, speech understanding, and voice and natural language information processing. Applications include manned and unmanned space systems, commercial aircraft, missiles, helicopters, marine systems, command/control/communications/intelligence systems, and manufacturing and process control systems.

Automated Data Systems (Degree in Computer Science, EE, Mathematics, or Physics)
Design, develop, implement, and maintain engineering computing business systems that include on-line application systems, project management, and hardware/software evaluation to meet company and/or customer requirements. Review systems changes to optimize company's business operations. Investigate and solve any problems in operational computing systems. Design, develop, implement, and maintain engineering test systems that include real-time applications systems for test control and data acquisition and analysis. Work includes hardware/software evaluation, design, and development to meet company and/or customer requirements. Investigate and solve any problems in operational computing systems.

Avionics System Design (Degree in EE, Computer Science, and/or Physics)
Avionics systems include; navigation, communication, utility, sensors and antennas, computing and data transmission networks, and operator displays and controls. An avionics system design engineer performs tactical and strategic avionics system preliminary design; assist in the specification, procurement and/or installation of major avionics systems from suppliers; and/or support subcontractor production schedules. Functions performed include resolution of technical problems; coordination with program and subcontract management; and development of design criteria, environmental standards, system performance evaluation, drawing release, system architecture, data processing and transfer, subsystem interfaces, and system level algorithms for management and control of avionic functions.

Base Construction Surveillance (Degree in CE, EE, or ME)
Analyze, check, coordinate, and report on missile base construction work performed by the subcontractors and associate contractors. This involves making calculations and completing required records and reports; conducting analyses and studies of construction and surveillance problems; remaining familiar with construction plans and drawings; compiling data and rough sketches of "as-built" drawings and changes to construction specifications, maintaining liaison among the customer, subcontractor, associate contractor, and supervisory and engineering personnel; and performing other surveillance checks and field engineering work assigned.

Base Installation Engineering (Degree in EE or ME, and related experience)
Assist in the installation, calibration, and checkout of functional equipment, launching-equipment sets, mobile launch equipment, manual missile equipment service equipment, and major assembly test sets. Includes scheduling, parts control, and production planning for this system installation.

CAD/CAM Systems (Degree in Math, CS, EE, Mechanical Engineering, or Manufacturing Engineering)
Evaluate, develop, and implement computer systems supporting Computer Aided Design and Manufacturing applications. Systems involve state-of-the-art graphics, complex numerical algorithms, and high-performance engineering work stations used to develop company product designs and to produce data to drive factory automation.

Ceramics (Degree in Inorganic Chemistry with ceramics experience for applied research and development positions. Degree in Ceramics, preferably at PhD level, for basic research positions)

Perform research and development on ceramic materials and processes required to solve high-temperature problems encountered in hypersonic aircraft and missiles. Work includes ceramic coatings, glasses, refractory oxides, carbides, silicides, borides, and heterogeneous systems such as cermets and fibrous materials. Emphasis is on ceramics suitable for elevated temperature structural application and new fabrication techniques resulting in more efficient structural designs. Conduct basic research studies of the factors influencing ductility.

Communications (Degree in EE, Math, or Physics at either bachelor or advanced degree level. Working knowledge of, or experience in, information theory or hardware design)

Perform the synthesis, analysis, conceptual study, specification, design, procurement, and test of communication subsystem hardware. Includes development of criteria, mathematical models, fast-time and real-time simulation, environmental analysis design standards, and system performance evaluation and test.

Computational Fluid Dynamics (Advanced degree in Applied Mathematics, AE, Physics, CS, or ME)

Develop and refine advanced computational methods for simulating detailed fluid flows such as occur in aerodynamics. Requires a breadth of high-level, multidisciplined scientific expertise encompassing the fields of advanced mathematics, theoretical fluid dynamics, and advanced scientific computer application. Entails (1) coupling of sound scientific principles with physical intuition to model physical flows and proper mathematical formulations, (2) innovative development, implementation, and evaluation of candidate solution algorithms, and (3) establishment of optimal problem and algorithm formulations with respect to computer system characteristics.

Computer System Analyst (Degree in Computer Science, Math, or EE)

Develop and evaluate numerical algorithms (including the formulation of mathematical models for these algorithms) required to satisfy specific performance or design requirements for operational computer hardware/software systems.

Computer System Engineer (Degree in CS, Math, EE, or Physics)

Analyze operational system requirements that result in the definition of computer system requirements. Develop plans and specifications for these computer systems. Perform trade and design studies necessary to accomplish preliminary design for computer hardware/software systems. Develop and implement computer programs for operational system simulators.

Computing Systems (Degree in AE, CE, EE, Math, ME, Physics or CS)

Evaluate, develop, and implement computing systems to fulfill company needs of various aerospace and airplane product requirements and to meet the needs of a variety of commercial and internal computing service customers. This includes the following areas:

- *Operating Systems*—Develop, implement, and monitor performance of large-scale computer operating systems and the integration of peripheral equipment, communication, and remote terminals.
- *Networking*—Specify, develop, implement, and maintain networking hardware and software supporting all levels of network communications across heterogeneous computing systems.
- *Languages*—Investigate, evaluate, and develop problem-solving language techniques and elements of the programming system including translation, assembly, and compiling.
- *Product Systems*—Develop real-time computing systems to monitor and control vehicle performance and to collect, evaluate, and display data for command and control systems. Develop and implement software systems for small and medium-sized computers used for on-line data acquisition and processing; commercial, military, and space systems simulation; and crew training simulators.
- *Equipment and System Evaluation*—Analyze and define the requirements for digital computing systems, specify configuration required, and evaluate present and proposed systems in a continuing program to improve the company's computing and communications facilities.
- *Product Support*—Provide generalized technical support for the computer center and programming community. Activities range from technical consultation on all services offered to new system and system component testing and certification. Additional responsibilities include systems change impact analysis and user coordination, system performance monitoring and reporting, and customer problem resolution.

Control Dynamics (Advanced degree in AE, EE, or ME and experience in feedback control systems)

Investigate (with analysis and computers) stability characteristics of flight vehicles and their control systems. Synthesize feedback control systems for stability augmentation and flight path control. Determine flight path control. Determine flight control system and servomechanism performance to tolerances. Verify automatic flight control system designs with analog and/or digital simulator. Analyze, synthesize, and evaluate inertial reference system and sensors such as gyroscopes, horizon scanners, and star trackers for application to the control of flight vehicles.

Control System Analysis (Advanced degree in AE, EE, or ME, with experience in feedback controls)

Perform synthesis, analysis, and test program definition of flight control systems for aerodynamic and surface vehicles, attitude control systems for space vehicles, and nonvehicular control systems such as precision pointing and tracking systems for sensors, antennas, and weapon-pointing systems. Includes determination of control subsystem requirements, functional design of system to meet requirements, and determination of software and hardware requirements compatible with the functional design. Work involves analysis of dynamics, hardware math modeling, automatic control theory and its application, simulation techniques, and developmental breadboard systems testing.

Control System Design (Advanced degree in EE, ME, or AE, with experience in feedback control systems)

Design and develop automatic control systems and control system components for aerodynamic, surface, and space vehicles and nonvehicular systems.

Countermeasures (Degree in EE or Physics, with applicable experience)

Analyze passive and active countermeasure and counter-countermeasure systems. Perform research in new techniques. Develop theoretical analysis and laboratory hardware for evaluation, testing, and effectiveness. Evaluate compatibility with ground, airborne, or space-based weapon systems. Develop both digital and hardware-in-the-loop simulators and simulation techniques. Countermeasure systems cover the frequency spectrum of acoustic, communication, microwave, millimeter wave, visible, and infrared devices. Requires knowledge of sonar, radar, electro-optical, and infrared sensor operation.

Crew Systems (Degree in AE, EE, IE, ME, Physics, or Psychology)

Perform analysis, development, design and testing of all interfaces between the man and the system, regardless of mission. It includes such systems as space systems, manned aircraft, missile systems, ground surface systems, and waterborne systems. The emphasis of the crew systems effort lies in the integration of the disciplines of human engineering (human factors), life support, displays and controls, design development, and testing of the total crew station. The overall goal of the crew systems effort is the effective and safe integration of the human component in the crew/machine system through the development and design of superior workplaces for the crew.

Customer Support Engineering (Degree in AE, CE, EE, or ME combined with experience on jet airplane systems)

Provide technical assistance to airline customers and field service engineers as required in the maintenance and operation of delivered aircraft. Provide ''quick-fix'' repair and design and expedite troubleshooting information. Participate in specialized problem consultations with the customer and, when necessary, make field visits to provide on-site recommendations. Involves engineering activity in the functional discipline areas of structures, mechanical systems, propulsion performance, and electrical/electronics engineering.

Data Reduction and Analysis (Degree in Math, Computer Science, EE or Physics)

Reduce or convert raw test data obtained from data recording systems to usable engineering quantities by means of automatic data reduction and computation equipment. Company equipment includes versatile data systems that record variables on disk or magnetic tape. Data from these machines is normally converted from raw to finished-time or frequency-domain form by use of digital, analog, or hybrid systems that are maintained for this purpose. As much use as possible is made of new techniques and instrumentation, and where necessary, new methods and equipment are devised.

Electrical/Electronic Circuit Design and Analysis (Degree in EE with special interest in circuit design and analysis)

Design and analysis of electronic circuitry associated with spacecraft, aircraft, missile, and ground system electronics, and of command, communication, guidance, navigation, radar, data processing, computers, control, checkout, and test systems. Includes design, analysis and/or application of microprocessors, analog and digital circuits, RF circuits, electronic power supplies, and fiberoptics.

Electrical/Electronic Installation Design (Degree in EE or ME)

Install and integrate electrical/electronics equipment and components into the airplane. This task includes the design and installation of equipment racks and panels and junction boxes and installation of exterior lighting, antennas, equipment cooling systems, and miscellaneous other electrical equipment systems.

Electrical/Electronic Measurement—Quality Control (Degree in EE)

Develop and implement research, development, and application of electrical/electronic calibration and inspection equipment systems, techniques and procedures, and analyze their performance for technical validity. Keep abreast of new equipment and techniques.

Electrical Systems and Equipment (Degree in EE)

Perform development work in electrical systems and equipment applications; analyze and synthesize electrical power generation systems from various energy sources; determine system dynamic performance as well as integration and compatibility with other systems; evaluate systems equipment to determine performance parameters; analyze and design distribution and protection systems and associated equipment as well as motor, air conditioning, and hardware; conduct research and development work in electrical power generation systems, system control and protection, and transmission utilization. Prepare systems and component specifications, procure equipment, and design and integrate the equipment into the vehicle or installations.

Electrical Wire and Component Installation (Degree in EE)

Perform design and development of wiring and electrical/electronic component integration and installation. Activities include determination of requirements, selection of wire and electrical components, integration of interconnect systems for wires, cables, connectors, junction boxes, switch panels and all components and subsystems of the electrical power and electronic systems.

Electrochemistry (Degree in Chemistry or Chem E, with experience in electrochemistry desirable)

Provide technical assistance to other engineering units; conduct research, development, and testing programs on electroplating, electrolytic finishing, electrochemical batteries, fuel cells, and other electrochemical processes. This involves studies of the effects of process variables upon structural properties of the base metals, development of new plating techniques, and parametric relationships for application of batteries and fuel cells.

Electromagnetic Compatibility (Degree in EE or Physics—BS or MS level)

Perform initial analysis of electrical/electronic equipment and systems to determine electromagnetic compatibility (EMC) design requirements. This includes interpreting or developing EMC specifications for specific equipment/systems and developing/applying EMC design requirements and engineering methods to equipment design and manufacture to ensure EMC at the system integration level. Plan and implement EMC specification tests and design/develop EMC test equipment to evaluate and verify equipment performance during qualification and system integration tests. Perform test data analyses and determine/implement corrective design.

Electromagnetic Pulse (Degree in EE or Physics)

Perform design analysis and testing to assess airborne and ground-based electrical/electronic systems relative to nuclear EMP and lightning phenomena. Analysis encompasses the use of computer code and mathematical models to determine the electromagnetic coupling paths and transfer functions from the external EMP environment to selected circuits within the systems. Perform laboratory testing and analysis at the component through subsystem level to determine upset and damage thresholds. Plan and perform system level EMP tests when system is exposed to pulsed and CW simulated fields. Provide design solutions to EMP-hardened systems and laboratory experiments to assess and verify hardening of systems.

Electronic/Electromagnetic Warfare (Degree in EE or Physics, with applicable experience)

Analyze, synthesize, develop, verify, and evaluate terrestrial, airborne, and space concepts, techniques, and subsystems for electronic countermeasure, electronic intelligence, and antisubmarine warfare including active measures (e.g., jamming, decoys, sonars), passive measures (e.g., information acquisition, confusion tactics, acoustics, magnetics), and all associated signal processing. Knowledge of command, control, communication, and exploitation is desirable.

Electronic Flight Controls (Degree in AE, EE, or ME)

Design and develop electronic control systems for engines and flight control for aircraft, surface and space vehicles, and nonvehicular systems. Includes transcription of functional requirements into hardware specifications, design, drawing release, supplier coordination, development of test plans and procedures, and system performance validation. Both digital and analog electronics are included. Electronics for some of the systems are designed and fabricated within the company. For this, preliminary design sketches are required, the most suitable components must be selected, and guidance of the procurement and testing of the hardware is necessary. Requires knowledge of automatic control theory and its application, electrical components, and electronic hardware and circuit design. Familiarity with hydraulic, pneumatic, and mechanical interfacing hardware is desirable.

Electronic Packaging (Degree in CE, EE or ME)

Perform design and development of the physical and mechanical aspects of electronic hardware for missile, spacecraft, aircraft, and ground support equipment. The activities involve printed-circuit board design; component selection; enclosure design; wiring, interconnect, and connector design, interface compatibility, shielding; and consideration of thermal, vibration, shock, humidity, and altitude requirements.

Electronic Parts Evaluation (Degree in EE with related experience)

Develop and document methods and procedures for determining electronic part failure rates and sampling techniques; provide technical support as required in the preparation of part application data, part procurement and process specification, and part failure catalog. Evaluate state-of-the-art microprocessors, LSI microcircuits and other parts for use in space, missile, and airplane equipment. Provide technical data to support design, parts procurement, and testing.

Electronic Test Equipment Design (Degree in EE)

Design and develop complex electronic test equipment associated with production and testing of advanced aircraft, missiles, and space vehicles. This involves using the latest electronic components and techniques to test both analog and digital systems. Tasks include analysis of system test requirements to define detail design of circuits; development testing of breadboard and preproduction models; performance testing; preparation of model specifications, and acceptance of test requirements.

Electronic Analysis (Degree in EE or ME)

Perform synthesis, analysis, and test management of electronics and electromechanical systems, including full electromagnetic spectrum coverage (e.g., infrared, radio frequency and audio), analog, digital, computation, sensing, processing, modulation, and demodulation. Includes development of criteria, mathematical models, fast-time and real-time simulation, environmental standards, and system performance evaluation utilizing laboratory, ground and flight tests.

Electronic Computer-Aided Design Engineering (Degree in EE or CS)

Analyze, design, develop, or acquire computing systems and software to assist engineers in the design, analysis, and test of digital and analog electronics. Includes stand-alone work station computing systems, large mainframe computers, applications software for design (schematic) capture, analysis (timing, simulation, circuit analysis, etc.), test requirements, data base management, communications, and networking.

Energy Conversion (Degree in EE, ME, or Physics)

Perform research on methods of conversion of solar, chemical, or nuclear energy to electricity. Applied research and development work are also being conducted on power conditioning devices for electrical power systems and regulation and control of the systems. Document the current and projected state of the art in power sources, energy conversion methods, and other projects.

Engineering Operations (Degree in Industrial Management, Industrial Engineering, Business Administration, or Business Law)

Develop and exercise the disciplines, methodologies, policies, procedures, and surveillance/audits of all items relative to engineering operations. These items include budget and schedule planning, work assignment, authorization planning and control, surveillance, and reporting. Also encompassed herein are the disciplines to ensure that the configuration and its interface requirements are defined, accounted for, and controlled, and that the parts release, drawing release, and drawing distribution systems are established and maintained. Also included is the establishment of disciplines concerning compliance and closeout of engineering contract talks for in-plant as well as supplier/subcontractor statements of work, together with applicable obligations of the Contract Data Requirements List. Further included are the disciplines relative to compliance with standard accepted formats for control of end item specifications. Engineering administration procedures and disciplines are also contained within Engineering Operations. All of the above items pertain to basic contract and change activities and to all types of acquisition contracts.

Environmental Control Commercial Airplanes (Degree in AE, CE, Chem E, EE, EM, and ME, Math, or Physics)

Design, develop and sustain environmental control systems used in commercial jet transport airplanes. This includes the following systems: air supply, which controls the flow, pressure, and temperature of engine bleed air delivered to the air conditioning system; air conditioning, which heats, cools, and controls the humidity of the supply air in response to temperature control system requirements; air distribution; cabin temperature control; cabin pressure control; air exhaust; electrical/electronic equipment cooling; protective systems such as wing thermal anti-icing, smoke detection/evacuation and duct leak detection; and electronic built-in test equipment.

Equipment Engineering and Management (Degree in EE, ME, or Chem E)

Assume responsibility for engineering associated with the development and acquisition of equipment for machining, chemical and thermal processing, bonding, forming, functional tests, and computer numerical control. This involves consulting with the in-plant requesting organization, preparing specifications, evaluating bid proposals and equipment, managing installation, and checking out procured equipment. Original design for special equipment application, modification, and retrofit of existing equipment, systems and processes with solid state controls, servo systems, advanced technology processes, miniprocessor and microprocessor controls are required. It includes remaining abreast of the latest advances in equipment for control design (trips of short duration to manufacturer's and user's plants are required) and the latest state-of-the-art developments regarding equipment application, machine processes, and computer systems.

Field Service Engineering—Aerospace (Degree in AE, CE, EE, ME, IE, and/or applicable experience)

Perform field engineering services to assist the product user to achieve and maintain proficiency in the operation and maintenance of products. Conduct analyses, investigations, and other technical tasks required to support field engineering services.

Field Service Engineering—Airplane (Degree in AE, CE, EE, or ME, combined with experience on multijet airplane systems)

Provide customers with technical assistance related to the operation and maintenance of airplanes. The primary responsibilities include; providing advice and recommendations to airlines involving problems encountered in operating or maintaining airplanes; details of any operations/maintenance problems encountered by the operator; providing counsel and training to assist airlines in their understanding and interpreting of drawings, design documents and service manuals; dissemination of information to customers relating to problems being experienced by other airlines operating the same equipment; and maintaining an awareness of the operators' dispatch reliability performance and, when necessary, making appropriate recommendations on ways of improving performance.

Flight Control Systems (Degree in AE, CE, EE, ME, Math, or Physics)

Analyze flight vehicles from the standpoint of aerodynamic stability and control. Establish stability and control criteria for piloted and nonpiloted vehicles through the use of wind tunnel test, numerical analysis, and analog simulation. Determine control concepts that will meet these criteria while causing a minimum penalty to vehicle performance. Based on established design criteria, define the type of actuation systems to be used on a given flight vehicle. Determine the display requirements necessary to manually control the design. Test all systems that are required to automatically control the path of the flight vehicle. Establish requirements to obtain stability and control data during flight test, and make adjustments or refinements in the system to achieve the desired flight characteristics as necessary.

Flight Mechanics (Degree in AE, EE, Math, ME, or Physics)

Develop trajectories of vehicles in space, planetary atmospheres, or water, based upon requirements to optimize performance in terms of vehicle weight, payload, end-point accuracy, minimum time, minimum energy, or other significant variables. Define and orient the equations of motion to the desired trajectory situation and to the depth required for conceptual, design, or operational trajectories. Find solutions for the trajectory equations through appropriate theoretical and numerical methods and develop mathematical solution tools as required for technology applications and advancement.

Flight Sciences (Degree in AE, Chem E, EE, ME, or Physics for applied research)

Perform fundamental research in aerodynamics, hydrodynamics, magneto fluid mechanics, two-phase fluid mechanics, subsonic and supersonic combustion, and flight mechanics. Investigations range from very slow subsonic flows to flow regimes substantially above the hypersonic region. Research is also being conducted in chemical kinetics and spectroscopy as a means of complementing the knowledge in the above-mentioned fields. In addition to theoretical and computational research, experimental investigations are conducted using ballistic ranges, hot shot tunnels, combustion-driven conventional shock tubes, electric-driven shock tubes, plasma jet-plasma accelerator combinations, low-density gas dynamics equipment, and other gas dynamics facilities for fluid mechanics work ranging from subsonics to hypersonics.

Flight Test Engineering (Degree in EE, ME, or AE, preferably with experience in aircraft or missile flight test planning and documentation)

Coordinate test requirements with the responsible project and assist in preparing missile or airplane flight plans. Plan and conduct flight tests for development and certification. Plan and implement data requirements, participate onboard by producing real-time data to determine quality and fulfillment of test requirements, and write test mission summary reports. Debrief and coordinate test results with project management.

Fluid Power (Hydraulics) Design (Degree in AE, CE, EE, or ME)

Design and develop hydraulic systems and components for aircraft, space vehicles, ships, ground transportation systems, and other vehicular and nonvehicular applications. Transcription of functional requirements into system configurations, design of hydraulic components, development of component specification, design of component installations, development of plumbing systems, responsibility for drawing release and configuration control, preparation of functional test requirements, customer and supplier coordination, and manufacturing liaison. Involves preparation of sketches, preliminary layouts, detail component layouts, installation layouts, writing specifications, reviewing proposals, selection of components, guidance of drafters, and verification of drawings. Computer-aided design (CAD) techniques are used extensively.

Functional Test Engineering (Degree in AE, EE, ME, or CS)

Provide for the functional test development, design, installation, and operation of instrumentation and data acquisition systems required to satisfy test and data requirements. Systems may include switching networks, simulators, and analog and digital interfaces. Includes analyzing state-of-the-art circuit designs and programming in high-level languages for testing of the designs. Also, development, implementation, and checkout of automatic test systems and testing stations, including operating system software development.

Gas Dynamics Research (Degree in ME, Physics, Physical Chemistry, or Aerodynamics, PhD degree preferred)

Conduct theoretical thermodynamic analyses involving steady and nonsteady gas mixture flow. This involves preparation of computational methods of analysis or reaction engine processes using digital computer programs. Conduct investigation in the general areas of combustion, detonation, shock dynamics, high temperatures, gaseous reactions, and magneto-hydrodynamics.

Geoastrophysics (Advanced degree in EE or Physics, preferably at PhD level)

Explore the nature of the environment that exists in the upper atmosphere of the earth and in interplanetary space. Particular subjects of interest are the earth's magnetic field, the ionosphere, radiation belts, cosmic rays, the moon and planets, the nature of interplanetary medium, and charged particles from solar flares. Both theoretical and experimental investigations are conducted. Experimental investigations embrace the techniques of radio propagation, including very low frequencies, solar radio astronomy, lunar temperature measurements in the infrared, and high altitude balloon and aircraft flights. Backgrounds in radio physics, upper atmosphere physics, cosmic rays, geomagnetism and astronomy are desired.

Ground Support and Handling Equipment (Degree in ME, CE, or EE)

Define concept and specify preparation, design, development and qualification testing of ground support and handling equipment associated with Company products. Ground support and handling equipment encompasses automated test equipment and related software; structural/mechanical equipment used to support, position, lift and transport; electrical/hydraulic/pneumatic equipment used to power, heat, cool, fill, flush and functionally test; tools and fixtures used to assemble/disassemble, repair, align, gauge, position, and adjust.

Ground Support Equipment (Degree in ME, CE, or EE, preferably with applicable experience)

Determine requirements for ground support facilities, equipment, special tools and devices required to service, maintain, repair, and test airplanes, engines, and components. Perform analyses of airplane configurations, systems, and components to ensure that equipment and facility requirements for all levels of maintenance are satisfied. Coordinate ground support equipment and facilities requirements with customers, subcontractors, vendors, technical organizations, and government agencies. Provide technical assistance to customers in their planning of ground support systems. Conduct surveys and perform audits of customer's total ground operations and maintenance tools, equipment, and facilities.

Ground Support Equipment Design (Degree in ME, EE, CE, or Industrial Technology)

Conceive, design, and develop mechanical or electrical ground support equipment such as special tools, test equipment, handling dollies, component installation and removal of fixtures/slings, component assembly/disassembly test hardware, etc., associated with the service, maintenance, and overhaul of commercial airplanes.

Ground Systems Design and Development (Degree in CE, EE, IE, ME, Math, or Physics)

Design and develop necessary ground electrical, electronic, and mechanical equipment; communication; logistic support and base support facilities; and related features comprising ground systems for guided missiles. This includes launch sequence, auxiliary power, automissile checkout, mobile launch site checkout, manual missile checkout, special sequence and monitoring, calibration certification equipment, ground instrumentation, and interconnecting cables and junctions.

Guidance, Tracking and Weapon Control (Degree in EE, ME or Physics, with experience in feedback controls)

Develop and synthesize potential guidance and weapon control systems as well as subsystem concepts for precision weapon delivery, fire control, aerospace vehicles, base defense, and vehicular defense systems. A spectrum of candidate guidance and weapon control system concepts is pursued based on mission requirements. Areas of technical knowledge are structured to meet specific mission/program requirements. They include; modern control theory, optimal estimation sensor computers, software systems design, pattern recognition and correlation theory, terminal guidance law formulation, dynamic system simulation, and guidance system testing and evaluation.

Heat Transfer—Spacecraft (Degree in AE, Chem E, CE (structures), EM, ME, or Physics)

Develop thermal management design approaches for controlling spacecraft temperatures. Employ computer analysis to define spacecraft thermal environment during ground operations, launch, and orbital mission phases. Develop analytical simulation models to predict thermal management system performance and to support spacecraft system design. Conduct tests in simulated space environment to verify thermal management system operation. Perform research to develop and demonstrate new thermal system hardware including heat pipes, cryogenic insulation, pumped fluid heat transport systems, cryogenic refrigerators, heaters, and durable surface coatings.

Human Factors (Degree in AE, EE, IE, ME, Physics, or Psychology, and Physiology with applicable experience)

Provide inputs and guidance to design organizations concerned with effective integration of man into current and future aircraft, space, and weapon systems. Analyze operational requirements and problems for effective human engineering solutions. Evaluate and conduct research on aircraft and support personnel, training equipment, control and display integration, and techniques and procedures to improve crew efficiency and flight safety. Guide and coordinate with design organizations to ensure effective interpretation of human factors into aircraft, space, and weapon systems design. Study military and industry efforts in defining human tolerances and stresses. Analyze instrument, equipment, and controls requirements for crewmen operating aircraft and other advanced vehicles.

Infrared Systems (Degree in EE or Physics with applicable experience)

Analyze and provide evaluation of infrared systems and components for surveillance, tracking, terminal guidance, and signature measurements. Conduct laboratory measurements of key components. Develop simulation techniques for infrared systems, targets, backgrounds, and atmospherics. Interface with subcontractors and government laboratories.

Instrumentation and Data Acquisition (Degree in AE, CE, EE, ME, or CS)

Design, install, calibrate, evaluate, and operate minicomputer-based instrumentation systems to record data for the support of test programs involving mechanical, propulsion, and structural systems as well as flight test programs. The test programs are related to research and development of aircraft, missile, and space vehicles and are conducted in laboratory environments such as propulsion test stands or tunnels, aerodynamic wind tunnels, space environmental facilities, and in-flight tests. Develop and implement software systems for small and medium size computers used for on-line data acquisition and processing in support of the above testing. Design, install, check out, and operate control systems, taking into consideration time phasing and specific response of the actuating systems. Integrate hybrid data acquisition systems to accommodate instrumentation packages and control systems, with the capability of "quick-look" verification of data against predicted reliability factors. In flight test or space programs, such data acquisition systems would be coupled with telemetry systems or other multichannel transmissions from remote locations.

Liaison Engineering (Degree in AE, CE, EE, or ME)

Analyze structural, functional and aerodynamic discrepancies (normally described on rejection tag), found on parts and assemblies, special repairs, customer pickups, and flight squawks. Search and evaluate supporting information and determine correct course of action. Perform review and audit of "Materials Review Board" activities for domestic and foreign supplies.

Life Support Environmental Control (Degree in CE, ME, Physics, or Physiology)
Research, design, and develop all aspects of the life support system that is developed to ensure the safety, comfort, and efficiency of the crew and passengers in all manned systems. The life support systems include all normal and emergency atmospheric control and supply, temperature and purity, inflight and ground escape, survival, seats, restraints, and personnel equipment, as well as toilets, galleys, and rest areas.

Logistics Engineering (Degree in AE, CE, EE, IE, or ME, preferably with applicable experience)
Analyze the intended operational system/equipment design to identify the human and material resources that are necessary to have the right items in the right place at the right time in operating condition to support a designated mission. Development and use of physical and mathematical models is entailed. Develop maintenance concepts that delineate maintenance support levels, repair policies, organizational responsibilities for maintenance, effectiveness measures, and maintenance environments. Develop logistic support requirements that identify maintenance tasks, task frequencies and times, personnel quantities and skill levels, test and support equipment, space/repair parts, and facilities. Coordinate maintenance program development with customers, vendors, and other company engineering organizations.

Logistics Research (Degree in AE, CE, EE, IE, ME, or CS, preferably with applicable experience)
Conduct research into logistics concepts and emerging technology for application to availability, maintainability, and reliability factors for system/equipment logistics improvement. Develop research projects that analyze requirements anticipated in the future, and implement research to derive a solution. Develop prototype hardware and software to increase product readiness and mission capability.

Management Control Information Services (Degree in Business Administration, preferably with engineering background)
Provide management with control information on the manner in which the engineering schedules and cost objectives are being met, and make recommendations, based on this information, to management for solving problems or improving company performance.

Management System and Consulting (Advanced degree in Business Administration)
Perform the analysis of company-wide business problems and recommend solutions to top management. In addition to serving in an internal consulting capacity, assistance is often required in the implementation of suggested solutions. Assignments may cover interdivisional problems, organizational realignments, and design of management control and information systems.

F-15 Eagle

Management Systems Simulation (Degree in AE, Computer Science, EE, IE, Math, ME, or Physics, preferably with advanced degree in Operations Research or Business Administration)

Provide the means to simulate business conditions in order to select and evaluate potential management decisions. This involves the application of advanced mathematical and scientific techniques to both management and operating systems. It includes problem definition, systems analysis, and the development and application of digital computing programs or systems. The scope of the work includes both engineering project support and general technology development.

Manufacturing Engineering (Degree in EE, ME, IE, General Engineering, Manufacturing Engineering, Production Technology, or Industrial Technology)

Plan the sequence of operations to fabricate a part or assembly in a manner that will efficiently utilize equipment, technical processes, and available skills. Perform test engineering that involves development of hardware and systems for mechanical, electrical, or electronic systems testing. Develop, document and implement preplanning and planning concepts for a variety of complex and technical situations, including the application of computer-aided design and manufacture. The manufacturing engineer analyzes engineering data, resolves producibility problems with design engineers, selects tooling, determines equipment specifications, defines test requirements, performs producibility reviews and analysis of new designs, and produces an efficient plan for manufacture of items. The manufacturing engineer works in a variety of technical areas including assembly, test, machining, welding, metalbond, fiberglass, plastics, composites, electronics, and computer graphics and programming.

Manufacturing Research and Development (Degree in Chem E, ME, EE, CS, or Met E)
Conduct applied research and testing for the development of new manufacturing processes, methods, and techniques, providing the technology required to manufacture new products. This activity includes development of the methods to adapt these manufacturing advances to current operations as well as new products.

- ✦ *Mechanical/Structural*—Conduct test programs related to sealing, metal bonding, finishing, welding, brazing, forming, plastics technology, advanced composite materials, and the use of structural fiberglass products.
- ✦ *Robotics*—Perform testing related to both electrical and mechanical fabrication and assembly, tool and end-effector development, and new applications research. Associated activities such as teaching and programming methods, vision systems, machine control and interfaces, and human interaction are also included.
- ✦ *Automated Storage and Handling*—Develop factory systems for automated transportation of materials, and their storage and retrieval. Associated activities include control systems, data collection/distribution, and machine interfacing.
- ✦ *Electronics*—Be knowledgeable in all phases of manufacturing technology related to the fabrication, assembly, and test of highly advanced electronic hardware. Printed wiring, microelectronics, assembly technology, wiring and cabling, and automated test technology are contained in this area of activity.
- ✦ *CAM/Computing*—Develop computer-aided manufacturing methods and integrate computing technology into the factory. This activity encompasses the design, development, and application of computing software and hardware solutions to manufacturing systems, and interfaces with the computer-aided design activity.

Materials Technology (Degree in Met E, Met, Chem E, Chem(MS), ME, or Min E)
Conduct engineering work in the wide range of materials and processes technologies applied to the design, development, and production of aircraft and aerospace products. The technologies are encompassed in two main categories.

- ✦ *Metals Technology*—Includes ferrous alloys, nonferrous alloys, corrosion and heat resistant alloys, and special metals and alloys, with associated metallurgy, heat treatment, metallurgical joining, mechanical joining, and beaming and bushing applications.
- ✦ *Chemical Technology*—Includes protective finishes, decorative finishes, sealants, fluids, lubricants, foams, rubber, fabrics, molded plastics, adhesive bonding, composite materials, and hybrid materials, with the full range of fabrication/processing techniques necessary for their applications. Primary activities are research and development, chemical and metallurgical analyses, specifications and standards controls, design coordination and production, and customer support. Complete and highly advanced materials laboratory facilities are available to support all activities.

Material Control (Degree in Business Administration or Industrial Technology)

Maintain control inventories for parts, materials, equipment, and supplies. Analyze material requirements and determine the scheduling of purchased items into the company and apply the necessary controls to ensure they arrive at their proper use point. Maintain balanced inventory levels so that there is neither a shortage nor an excess condition.

Mechanical Flight Control Design (Degree in ME, AE, CE, EM, Math, or Physics)

Design and develop flight control systems, subsystems, components, and installations for commercial and military aircraft and space vehicles. The task consists of translation of functional aerodynamic or other control requirements into control system configurations, encompassing most areas of the vehicle and involving cable systems (cranks, quadrants), linkages, gears, nobacks, clutches, brakes, and other similar mechanical devices. From preliminary sketches, analysis and layouts, the job leads through geometric and kinematic analysis to preparation of specifications, preparation of component and installation layouts, vendor selection, vendor coordination, drafter guidance, drawing verification, functional test definition, and manufacturing liaison. Computer-aided design (CAD) techniques are used extensively.

Mechanical System Engineering (Degree in AE, CE, Chem E, EE, EM ME, Math, or Physics)

Perform research, design, development, and testing of mechanical systems used in aircraft, guided missiles, and spacecraft. This includes the following types of systems: personnel environment (all aspects of air conditioning, including heating, cooling ventilation, pressurization, air purification, and humidity control); equipment conditioning (cooling, heating, pressurizing, or otherwise air-conditioning electronic and other components to overcome conditions created by high-speed vehicle performance); surface conditioning (deicing, anti-icing, cooling, defrosting, and rain removal); hydraulics (hydraulic systems and hydraulically actuated mechanisms, such as brake flaps, spoilers, and landing gears); pneumatic ducts, pneumatic manifolds and turbo units; controls (ailerons, elevators, rudders, and engine controls); and mechanisms (mechanically operated systems such as transmissions, rotor hubs, stabilizer jackscrews, and flap drive mechanisms).

Metals Engineering (Degree in Met E or Met for applied research PhD in Physical Metallurgy for basic research)

Perform research or development work in the application and properties of alloys, metal-forming techniques, welding, brazing, and heat treating for aircraft, satellites, and space vehicles. Assist in the selection of alloys and conduct failure analysis.

Microcircuit Development (Degree in EE with related experience)

Conduct engineering studies on design applications for integrated microcircuits. This includes development of electronic equipment design standards and an experimental model of equipment utilizing integrated microcircuits.

Microwave Circuit and Subsystem Technology (B.S., M.S., or PhD in EE)

Analyze, design and test RF, microwave, and millimeter wave components, circuits, and subsystems. Responsibilities include the use of computer aid to design, development of breadboards and preproduction models in the laboratory, and investigation of new millimeter wave components and circuit techniques. Work is conducted to meet requirements of future communications, electronic countermeasures, and radar systems.

Missile and Space Vehicle Propulsion (Degree in AE or ME)

Perform analytical investigations in support of preliminary design studies of missiles and space vehicles and plan experimental testing on solid and liquid propellant rockets, nuclear, electrical, ramjet, and other advanced propulsion systems; evaluate the application of all types of propulsion systems and supply performance and installation information to design groups. Experimental work could involve the development of solid and liquid propellant rocket systems, establishment of propellant stability criteria, liquid propellant slosh and impedance characteristics, solid propellant burning characteristics, research in and development of reaction controls, rocket thrust vector controls, thrust termination devices, and propellant expulsion systems.

Navigation Systems (Degree in AE, EE, ME, or Physics with experience in feedback controls)

Perform synthesis, analysis, design and evaluation of inertial navigation, radio navigation, hybrid navigation systems, radio command guidance systems, or inertial components in the development of navigation and guidance systems for aircraft, missiles, space, and surface vehicles. Includes the transcription of system functional requirements into hardware specification and design, drawing release, supplier coordination, test planning, and system performance validation. Areas of technical knowledge include modern control theory, optimal estimation, inertial sensor analysis and design, system dynamics, communication theory, inertial system testing, and evaluation.

Noise Technology (Degree in Engineering of Physics)

Support new business and current production programs in noise technology, including the development of acoustic lining, jet sound suppressors, cabin noise insulation, and structural tuning/damping. Estimate community, ramp, and interior noise. Assess impact of existing and proposed noise regulations. Advance the state of the art through research and test programs. Participate in technical discussion and design reviews with manufacturers, customers, and concerned engineering and marketing organizations. Maintain technical liaison with government agencies, technical institutions, and technically related scientific and industry organizations. Provide related laboratory support.

Nuclear Physics and Engineering (Degree in Physics, Nuclear Engineering, or ME)

Perform applied research in such areas as nuclear radiation physics, nuclear radiation effects involving the detonation of nuclear weapons, operation of nuclear power plants, and protection of the crews and equipment in flight vehicles from these effects. Analyze nuclear heat transfer problems in high-speed flight vehicles. Investigate company radiation measurement requirements. This involves providing consulting services on radiation measurement techniques, determining measurement accuracy, and designing measurement devices. Perform analytical and experimental research on the theory, performance, and application of advanced nuclear reactor concepts that could be useful for propulsion. This includes techniques for using reactor heat directly for propulsion.

Operations Analysis/Systems Analysis (Advanced degree in Engineering, Math, Operations Research, or Physics)

Apply engineering and mathematical methods to study the requirements of potential customers or consumers, and the systems, actual or postulated, intended to serve these requirements. The activity includes: derivation of customer requirements; mission analysis; determination of the effectiveness of systems and combinations of systems; research on system utilization modes and tactics; analysis of systems effective sensitivity to changes in system characteristics and utilization modes and tactics, methodology research and development; development of figures-of-merit; development of special factors including political, ideological and economic considerations, as well as scenarios and threat definition; and economic analysis. Appropriate qualifications include knowledge over a wide range of systems performance capabilities and vulnerabilities, understanding of military and consumer operations, and ability to mathematically analyze and simulate complex systems utilizations and interactions. Outputs include statements of mission or customer essential needs, evaluations of system effectiveness, models for analyzing and evaluating systems, description of suitable and effective utilization modes, and descriptions of appropriate special factors.

Operations Research (Degree in Business Administration, and advanced degree in Mathematics, Operations Research, Physics, or IE with experience or training in Operations Research)

Apply mathematical concepts and techniques to the analysis of business problems and to the decisions process. Applications include inventory control, production scheduling, estimating and forecasting, resources allocation, design of experiments, and the determination of requirements for management control systems. Increasingly, operations research techniques are being applied to military and space system preliminary design, analysis, and optimization. In the determination of overall cost effectiveness, maintenance policy and procedures, and reliability budgets and goals, the techniques of linear programming, game theory, stochastic processes, and inventory analysis are applicable and useful. Other assignments involve the design and use of mathematical and symbolic models, simulations, and both linear and dynamic programs.

Operations Research/Systems Analysis (Degree in AE, CE, CS, EE, ME, Systems Engineering, Operations Research, Mathematics, or the Sciences Advanced Business Administration Degrees)

Apply engineering, scientific, and mathematical methods to study the capabilities, survivability, and effectiveness of military and space systems in the performance of their missions. Activities include system cost-effectiveness optimization, research on system capability improvement potentials, development of new employment modes and tactics, derivation of customer requirements for new systems, model development, and development of new analysis methodology. Analysis projects often require special consideration of political, ideological, and economic factors and often lead to involvement with problems and issues at the forefront of contemporary geopolitical and strategic thought.

Passenger/Cargo Compartment Design (Degree in AE, CE, or ME)

Develop, design, and release to manufacturing those drawings and specifications required to establish the interior configuration of vehicles to satisfy the requirements of passenger comfort, appeal, and service, as well as meet customer requirements. In the accomplishment of these activities, attention will be given to such elements as sidewall and ceiling panels, insulation, stowage units, lavatories, emergency equipment, emergency lighting, escape slides, communication systems, oxygen, water, galley and waste systems, general cabin layout, and baggage and cargo handling systems. Design analysis will include structural integrity of metallic and plastic materials, mechanisms, incompressible and compressible fluid mechanics, clerical equipment, and manufacturing processes.

Performance Analysis (Degree in AE)

Predict flight vehicle performance and variations affected by aerodynamic configuration, propulsion system, and mode of operation. Interpret and apply given propulsion and weight data. Determine the relative effects of major design parameters and their optimum relationship to each other to guide the design improvement of existing or projected vehicles designs. Calculate the specific performance characteristics of designated vehicles from the point of takeoff or launch, through acceleration, cruise, reentry, or deceleration and landing. Certify airplanes and evaluate flight data.

Personnel Safety Engineering (Degree in Chem E, CE, EE, ME, Met E, Min E, Chemistry, or Physics)

Perform personnel safety analysis, which encompasses design, test, and operational activities. These analyses are conducted in concert with the weapons systems safety analyses and identify both personnel and operational equipment hazards. Develop safety requirements documentation which is task unique and program oriented. These documents are based on engineering drawings and procedures and the manufacturing sequence. Conduct safety analyses on equipment, tooling, and facilities that identify design requirements that must be included in the drawing system. Review and approve, for safety, A&E type drawings; transportation and handling; material handling equipment and procedures; and be familiar with ICC regulations, ANSI Standards, and applicable laws and contractual regulations.

Plastics (Degree in Chemistry or Chem E)

Perform research and development work on plastic materials and processes for missiles, spacecraft, and aircraft, including the high-temperature structural applications for these materials and the new fabrication techniques that will result in improved structural design. Study ablation materials for vehicles reentering the earth's atmosphere.

Product Assurance Engineering (Degree in AE, CE, EE, IE, or ME, preferably with applicable experience)

Be responsible for system reliability and maintainability performance parameters from system concept through design, test, manufacturing, and field use. System concept definition requires technology evaluation, trade study effort, and math modeling to define system parameters. Design support effort requires definition and evaluation of component stress margins, Built-In-Test mechanization, failure modes and effects, failure and maintenance rates, and subcontractor design review. Test phase effort requires definition, monitor, and evaluation of statistically defined tests to demonstrate requirement achievement. Manufacturing effort requires definition and monitoring of effective product screens to detect and eliminate marginal components and manufacturing defect escapes; this phase also requires physics of failure analysis and corrective action monitoring. Throughout test, manufacturing, and field use, a failure data bank is maintained for identification of problems, trends, and corrective action effectiveness. Product assurance engineers coordinate directly with the project design and technology organization personnel and must be technically qualified to communicate effectively with them on an engineer-to-engineer basis. Product assurance engineers critique features of vendor-designed, as well as company-designed, equipment for compliance with established reliability and maintainability requirements. They also closely follow the experience of the aircraft in airline service to evaluate and document this experience of the aircraft in airline service to evaluate and document this experience with the objective of applying lessons learned to the improvement of existing and future airplane models. The determination of failure modes, the impact of single and multiple failures on system and airplane operations in service, and the determination of cost of ownership of competitive design options are frequently requested studies used in fixing final design configuration. Product assurance engineers must be capable of effectively utilizing and directing the efforts of nonprofessional personnel for routine data searches and preparation of graphic or tabular data. A knowledge of computer technology is also a requirement.

Product System Engineering (Degree in AE, CE, EE, ME, or Physics)

Apply system engineering techniques to the development of system/subsystem requirements and the definition of the system. This includes conduct of system requirements analysis leading to subsystem hardware and software performance, design, and verification requirements/specifications for operation, maintenance, assembly, and checkout of the system and definition of functional, physical, and procedural interfaces between subsystems and elements. Incorporate requirements for safety, human factors, survivability, logistics, and cost. Develop other required subsystem integration data. The skill requires a broad knowledge in one or more technical fields, associated end item hardware, system engineering practices, and a thorough understanding of specification preparation. Outputs include documentation of functional flow diagrams, requirement allocations, subsystem trade study results, development specifications, integration data including interface definition, system description, design reviews, and verification of compliance to specifications.

Product Development (Degree in AE, CE, EE, EM, or ME)

Address those areas that potentially could bring new business into the company. Do so primarily by applying new concepts and/or technologies to meet the needs of the customer. This requires a broad spectrum of disciplines for the design, development, analysis, and assessment of the proposed new product to ensure that it meets customer needs and company objectives. Technical skills and the ability to effectively communicate the results of technical studies are required.

Program Planning (Degree in Manufacturing Engineering, IE, or Business Administration; MBA with engineering background for latter degree)

Develop initial manufacturing program plans on new company business. Develop long-range program schedules and analyze related space and facilities requirements to support such programs. This involves close liaison and contact with the preliminary design and other engineering and operation organizations in order to obtain and analyze specifications, sketches, and other engineering data necessary for preparing plans to accomplish the program, and evaluating the impact of these plans on the operations and objectives of the manufacturing organization. Other program planning positions involve using electronic data processing equipment to obtain schedules, scheduling status on company product programs, and performing detailed planning for evaluation of these programs.

Propulsion (Missile/Space) (Degree in AE, CE, EE, EM, ME, Chem, Chem E, or Physics)

- ✦ *Propulsion System Design*—Establish design requirements and objectives, make design sketches and layouts, prepare specifications for purchased equipment, and make and release production drawings and documents for missile and space propulsion system installations, including rockets, ramjets, and advanced propulsion concepts. Perform supporting trades. Integrate and provide compatibility for various system requirements. Assume responsibility for integrity and quality of delivered propulsion systems and their installations.
- ✦ *Propulsion System Analysis and Test*—Calculate expected performance of missile and space propulsion systems and components statically and over a range of flight conditions. Plan and execute research and development of missile propulsion systems, engine installations, air inlets, nozzles, thrust vector control, fuel management, and fuel system components, to verify calculations and show compliance with specification. Conduct research and development in internal ballistics, propellant feed systems, nozzle and exhaust plum definition, cryogenics, and other thermodynamic and fluid dynamic studies. Develop the necessary analytical tools and assemble the required technology data base.
- ✦ *Propellant Research*—Investigate chemical and physical properties of liquid and solid propellants, including high-energy and toxic materials handling and testing, and solid propellant technology and inspection techniques. Conduct theoretical analyses on propellants of current or future interest for missiles or space vehicle propulsion systems, which may include cryogenic fuels, fuels built around metastable states, or plasma propellants.

Propulsion (Aircraft) (Degree in AE, CE, EE, EM or ME)

- *Engine Analysis*—Participate in the development of engines by extensively analyzing engine thermodynamic cycles and requirements, carrying out technical discussions with engine manufacturers and operators, ground and flight testing engines, and executing engine specifications for procurement.
- *Power Plant Design*—Establish design requirements and objectives; prepare specifications for specially purchased equipment; make preliminary design sketches and layouts; coordinate with other projects, research groups, flight test, and manufacturer's representatives; and make and release final drawings on the following: power plant installations, accessories, nacelles, acoustic treatment, and thrust reversers, fuel and tank vent systems APU systems, air refueling and tank purging systems, and fuel tank installations.
- *Propulsion System Component Analysis*—Conduct analyses to establish the design and performance of aircraft engine installation components. These include: fluid mechanics of inlets, ducts, nozzles, and thrust reversers; heat transfer related to system cooling and engine environment; and control of engine and thrust reverser.
- *Engine Propulsion Test*—Plan and execute tests of jet and gas turbine engines, associated fuel systems tanks, and all components affecting the propulsion system, and analyze the results of these tests to try to improve performance or correct operational problems involved with the systems. This involves checking pressure, flows, temperature, and other phenomena with the aid of extensive electronic instrumentation, and testing engine inlets and nozzles at subsonic, supersonic, and hypersonic speed with company wind tunnel facilities.
- *Propulsion Control System*—Perform synthesis, analysis, and test program definition of control systems for engine inlets, nozzles, and their integration with gas generator and airplane controls. This includes determination of subsystem requirements, functional design to meet requirements, and determination of software and hardware requirements compatible with functional design. Requires experience in dynamics, hardware, math modeling, digital control theory and its application, analog and digital simulation techniques, and developmental breadboard systems testing.
- *Propulsion Fluid Dynamics*—Develop and apply computer programs for analysis of the flow phenomena characteristic of propulsion installations. This includes preparation of analytic geometry descriptions, generation of computational meshes, numerical solution of partial differential equations, and turbulence modeling of two and three dimensional viscous flows in flight regimes from subsonic through hypersonic.

Quality Control Research Engineering (Degree in Chem E, EE, ME, Met E, Math, or Physics)

Research and develop new techniques and instruments for the performance of product inspection, laboratory analysis, and nondestructive material testing. Refine such techniques and instruments, directing their application to testing plans designed to prove conformance of product or materials to established quality specifications. Develop and apply statistical techniques to various quality control functions such as sampling plans in inspection of parts, analysis of data in research projects, and the controlling of a manufacturing process to an established specification. Develop and direct the use of measurement techniques and standards in the physical-electrical measurement field such as pressure, vacuum, force, dimension, flow, viscosity, acceleration, voltage, current, resistance, and power. Provide technical direction in all matters concerning the quality of materials, systems, and processes used.

Radar (Degree in EE)

Perform theoretical and experimental studies to advance the state of the art in active and passive radar and microwave systems. Involves analysis, synthesis, development, verification, and evaluation of radar concepts, techniques, systems, subsystems, circuits, microwave RF and IF components, and all associated signal processing, antenna applications, and computer analysis. Verification and evaluation includes simulation, breadboarding, brassboarding, and laboratory, field, and flight testing. This research covers the spectrum from very low frequency through millimeter waves.

Radar and Beaconry (Degree in EE with applicable experience)

Perform theoretical and analytical studies to improve the design of future radar systems; apply information theory in the analysis of heavy radar design; apply electromagnetic wave theory to radar antenna design and propagation of radar signals; design and develop pulse circuit data link systems, beaconry systems, and microwave equipment; analyze systems performance and work with radar systems applications.

Radar Cross Section/Propagation (Degree in EE)

Conduct theoretical and empirical analyses to support the design and development of aircraft and missile radar cross section. This includes dielectric and magnetic absorber materials and structure research to reduce specular reflections, diffraction effects, and traveling/surface wave backscatter; study of shaping techniques and low signature sensor designs; and analysis of synergistic methods for integration of passible signature controls with active electronic countermeasures. Perform analyses to determine electromagnetic propagation characteristics through a variety of media including atmosphere, earth, plasma, nuclear environments, and stratified dielectric and/or magnetic materials. Determine the propagation properties for communication, data, and radar system links.

Radiation Effects (Degree in EE or Physics)

Perform analysis, design, and evaluation of systems exposed to the radiation environments produced from the detonation of nuclear weapons or from space environments. Transcribe system requirements into hardware specifications and design, coordinate supplier activities, and assist in drawing release. Assess the impact of transient upset and permanent damage on system operation. Plan and perform radiation environment testing for design development and qualification.

Real-Time Simulation (Degree in AE, CS, EE, Math, ME, or Physics)

Develop real-time digital computational simulations. Solve engineering problems such as feedback control systems, heat transfer, and aerodynamic stability. Design and develop flight simulation equipment incorporating analog and combined analog/digital computers. Specify and develop computer equipment to satisfy computer technical requirements.

Real-Time System/Simulation Software Engineering (Degree in CS or EE)

Design, develop, integrate, and test software, special-purpose simulation computers, and/or end-item computers for real-time aircraft, aerospace, and special-purpose systems. This includes the engineering design and integration of the computer software system with simulator, airborne, and/or special-purpose hardware.

Servomechanisms (Degree in EE or ME)

Analyze, synthesize, test, evaluate, and develop electromechanical and hydro-mechanical servomechanisms. This includes systems to operate landing gears, steering, brakes, spoilers, and control surfaces on aircraft, rocket nozzle actuators to control thrust direction, aerodynamic control surfaces or antennas on guided missiles or space vehicles, and reaction control systems for application to space vehicle attitude stabilization.

Software Development Engineering (Degree in CS, Math, EE or Physics)

Design, develop, implement and test computer programs for mission-critical computing hardware/software systems embedded in larger operational systems, the primary purpose of which is not data processing. Implement components of systems on digital computers. Develop and implement programs to assist in designing, simulating, debugging, and testing of operational software systems. Work could entail development using host systems, then downloading object programs and testing them on target systems. Applications include simulators, automatic navigation, guidance and control, satellite communications, command control/communications/intelligence, energy management and supervisory systems, and many other special-purpose or real-time applications.

Software Research Engineering (Degree in CS, Math, or EE)

Plan and conduct research and development programs on software development methodology techniques and automated tools to reduce the cost and improve the quality of software developed for military and space systems. Acquire, document, and disseminate knowledge of software engineering and its applications, software systems design, system simulation, and computer programming. All facets of problem solving, from concept definition through implementation, are involved.

Software System Engineering (Degree in EE, Physics, Math, or CS, advanced degree in EE or CS in beneficial)

Analyze functional, performance, and human interface requirements for mission-critical computing hardware/software systems embedded in larger operational systems, the primary purpose of which is not data processing. Allocate system requirements to hardware, firmware and software end-items and modules. Support the development of system architectures and develop software architectures using advanced system analysis and requirements traceability tools. Perform system-level analysis and trade studies to support selected design approaches. Support proposals, projects and/or research activities. Individual applications may include process-control systems, commercial or military aircraft, marine systems, manufacturing systems, missiles, manned and unmanned space systems, helicopters, and others.

Space Vehicle and Missile Guidance and Tracking (Degree in EE, AE, Math, or Physics)

Perform research, design, and development for space vehicle guidance, missile guidance, orbit determination, and trajectory estimation systems. Apply modern control, optimization, and estimation theory; develop techniques for spacecraft rendezvous and docking; and design algorithms suitable for flight computer implementation. Translate system requirements into guidance subsystem design and perform engineering trades. Develop analytical computer models and digital system simulations. Applicable technical disciplines include optimal control and estimation, orbit mechanics, numerical methods and statistics, computer modeling, and software design.

Spares Engineering (Degree in AE, CE, EE, or ME, plus related aircraft industry or military experience)

Review engineering design drawings to determine potential spare parts. Analyze potential spares and develop initial and sustaining technical provisioning data, including parts interchangeabilities. Using spares technical data, develop discrete spare parts recommendations customized to a specific operator's or military user's requirements. Make presentations and participate in meetings with airlines and military customers locally or at their facilities worldwide regarding initial provisioning, sustaining spares support, or assisting in sales campaigns.

Standards Engineering (Degree in CE, EE, or ME)

Develop, implement, and maintain company standardization of parts, tools, equipment, materials, finishes, design applications, interchangeability, and availability. Review new standards released by the military services for their effect on company standardization programs. Investigate and help solve any problems resulting from changes in the company standards programs. Review surveys by government standards agencies, and comments of other companies and organizations on these surveys, to help prepare comments or recommendations.

Stress Analysis (Degree in AE, CE, EM, or ME)

Establish structural design criteria to ensure that lightweight, economical design is maintained in aircraft, surface craft, missiles, launch vehicles, or spacecraft without sacrificing structural integrity or fatigue resistance. Determine the magnitude and distribution of vehicle maneuvering, gust and impact loads for all operating regimes—both the subsonic and supersonic on land, sea, air, or outer space, if applicable. Determine requirements for and initiate laboratory or flight testing of vehicle and structural components, and evaluate test results and application to the vehicle design. Perform strength, fatigue, residual strength, and survivability analyses of detail design of structural components. Approve production design drawings. Coordinate with engineering and manufacturing groups and with the customer and appropriate government agencies during production and throughout the service life of the vehicle. (Accident and service structural investigations comprise an important part of this work.) Perform research in such fields as composite structure development, subsonic to hypersonic airload prediction, methods of stress analysis, blast effects, thermal effects of supersonic flight or space radiation, fatigue, and fracture mechanics.

Structural Design (Degree in AE, CE, EM, ME, or Industrial Technology)

Assist in establishing design requirements, criteria, and objectives. Design pressurized and unpressurized structures for aircraft components such as wing, fuselage, empennage, and landing gear. Designs involve state-of-the-art materials and processes such as advanced composites and metallic and nonmetallic bonding, including honeycomb applications. Design activities involve both static and fatigue analysis, layout preparation, and coordination and release of final production drawings. CAD/CAM methods are employed to the greatest extent possible. The design process includes extensive coordination with the supporting technical staff organizations, manufacturing, and tooling. Estimating, scheduling, and reporting are an integral part of the design task. Producibility of the designs is of utmost importance.

Structural Dynamics (Degree in AE, CE, EE, EM, or ME)

Establish requirements to prevent dynamic instability (such as flutter, vibration, and landing gear shimmy) in subsonic to hypersonic aircraft, missiles, and space vehicles. Conduct analyses and furnish data to other staffs on transient vibratory response and dynamic load increments on flight vehicle structures. Furnish information to project design groups on the intensity of environmental vibration throughout a vehicle under various operating conditions, and establish requirements for vibration qualification tests of airborne equipment. Conduct wind tunnel tests of scaled elastic flutter models, evaluate data and make suitable recommendations to project or preliminary design personnel; conduct flight flutter tests of full-scale aircraft in order to verify that margins of safety with respect to flutter are adequate.

Structural Testing (Degree in AE, CE, EE, EM, or ME)

Conduct tests to verify the structural integrity of complete aircraft, missiles, and space vehicles as well as structural components, assemblies, and systems. Establish, monitor, and analyze all wind tunnel, structural and flight tests required to demonstrate the structural integrity of a vehicle or component. Included are tests to verify existing designs and developmental tests to improve future designs of manned and unmanned vehicles; testing to point of destruction of major components for the purpose of determining the ultimate strength and durability of the structure for flight and ground handling conditions that govern vehicle design; vibration tests of components and entire vehicle structures; flutter tests of models and flight vehicles; sonic fatigue testing of high-noise-level effects on structural materials and components; tests to determine the effects of the operational environment temperature, pressure, and moisture; and associated research on supersonic and hypersonic aircraft, missiles, and space vehicles.

Support Equipment and Facilities (Degree in AE, CE, EE, ME, IE, and/or applicable experience)

Identify, evaluate, and design ground support facilities, equipment, tools, and devices required by the product user to maintain, overhaul, modify, and test the product. Develop equipment and facilities to improve product operation site effectiveness.

System Configuration Development (Degree in AE, Chem E, CE, EE, ME, Physics, or Math)

Perform the system configuration design evolution, from definition of the initial concept through successive iterations to a preliminary design suitable for detail design. Be technically familiar with the impact on the system configuration of performance requirements, subsystem technology, operational requirements, and maintenance requirements. Conduct trade studies that consider physical and functional integration of the subsystems into the configuration and the configuration into the total system. Provide visibility of the applicable system configuration design, and control its evolution through preparation of inboard/outboard profiles, general arrangement drawings, base layouts, launch control center layouts, and system configuration description documentation.

System Integration (Degree in EE and applicable experience)

Establish standards for equipment to fulfill the requirements for aircraft and missile electronic equipment. Establish programs to ensure compatibility of all associate or subcontractors' electronic equipment. Establish requirements for electronic integration during base system integration tests. Establish reliability requirements for electronic design, and monitor reliability factors during development and testing of systems equipment. Handle engineering liaison on electronic equipment between manufacturing and test areas.

System Engineering—General (Degree in AE, CE, EE, or ME and training in the Systems Engineering discipline)

Develop and deploy a system that: (a) transforms an operational need into system performance requirements and a system concept optimized through the use of an iterative process of definition, synthesis, analysis, design, test, and evaluation; (b) integrates related technical requirements and assures compatibility of all physical, functional, and program interfaces and constraints in a manner that optimizes the total system definition and design; (c) integrates the requirements of reliability, maintainability, safety, survivability, human factors, etc., into the total system capability; and (d) assures the verification of compliance with system requirements. Apply standards and configuration control in the course of development, and ensure the delivery of all system engineering documentation in support of development of the system.

System Safety (Degree in AE, Chem E, CE, EE, Physics, or Math)

Apply scientific and engineering principles to accident risk management. Identify system hazards and institute action to prevent accidents and minimize risk. Provide for the safety of personnel using the system and of the public, and for the protection of equipment included in the system. Beginning with the conceptual phase of the system, identify and control hazards as the design progresses, and continue the process through the entire life cycle. System safety engineers require a broad knowledge of safety standards, both military and civilian, knowledge of safety analyses techniques, and the ability to understand and analyze new systems and operational concepts. Develop detailed qualitative and quantitative analyses to provide a logical evolution of safety requirements applicable to design, development, test, production, servicing, modification, refurbishment, maintenance, handling, transportation, training, disposal, deployment, and contingency operations. Establish safety goals, interpret safety standards, participate in design and operational reviews, and provide liaison, safety surveillance, and records experience data for use on future programs. Provide support for use in management of accident risk caused by human error, environment, deficiency or inadequacy of design, and component malfunction or interaction.

System Test Engineering (Degree in AE, ME, EE, Math, or Physics with aircraft or missile systems test experience)

Provide technical assistance in functional testing and processing of aircraft, guided missile, and advanced vehicle component systems. Perform system and subsystem testing of entire weapon system complexes; analyze the test results for malfunctions or deficiencies and initiate corrective action including the coordination of test results with design engineers. Monitor revision of test requirements, test design, and procedures that result from system design changes.

System Training (Degree in AE, CE, EE or ME, interest or experience in teaching preparation of courseware)

Develop instructional materials from engineering data for presentation to customers and company personnel. These materials cover the operation and maintenance of military, other government, and commercial systems. Apply the Instructional System Development (ISD) approach to the preparation of course materials for traditional classroom and automated instruction.

System Synthesis/System Design Analysis (Degree in AE, EE, ME, Systems Engineering, or Physics)

Apply engineering methods to establish system performance requirements in response to mission or essential needs; to synthesize the system; and to describe its operational use. Postulate alternate system approaches. Perform mission analyses and understand the interrelationships between mission requirements, system characteristics, and system effectiveness using predictive models as appropriate. Establish performance, capacity, and response requirements. Synthesize and optimize system concepts considering technical feasibility, performance effectiveness, need for verification, operational suitability, life cycle costs, and growth capability. Assess risks in performance and costs and predict optimal growth paths. This skill requires a broad knowledge of military missions, current systems and their operation, state of the art knowledge in key subsystems, an ability to formulate new systems or operational concepts, and an ability to analyze their effectiveness. Generate a description of the system and its operational concept; a statement as an entity; allocation of requirements to the major functional areas; and a definition of the functional area interfaces.

Technical Publications and Training Engineering (Degree in AE, CE, EE, or ME, additional education or experience in writing desired)

Develop and present engineering data in clear and concise written form for operation, flight, maintenance, and overhaul manuals. Research technical information, create detailed instructions using a computer work station and computerized data bases, determine needed illustrations, and validate data using the actual hardware. Use one-half of work time for research and coordination, one-half for writing. Prepare and present material for instruction in commercial and military aircraft and missiles.

Telecommunications (Degree in CS, EE, or Math, with experience of interest in Communications Theory, Networks, and Digital Switching Technology)

Perform analyses, specification, design, procurement, integration, and test of communication systems. Develop modeling and simulation tools, system performance criteria, and monitoring techniques system engineering, and integration.

Telemetry (Degree in EE and/or applicable experience)

Design, develop, and validate the performance of aerospace telemetry systems that are used to obtain test and performance data on missiles, spacecraft, satellites, marine craft, and other products. Covers all aspects of telemetry, from transducers or sensors through signal conditioners, multiplexers, RF transmitters and receivers, demultiplexers, and digital interface equipment to enable computer processing of test data. Have knowledge of latest sensing techniques, communication theory, multiplexing techniques, and digital logic design. Assignments vary from specification, analysis, and evaluation of a telemetry system in order to meet specific program requirements to conceptual development of new hardware through breadboarding, modeling, and packaging design.

Test Engineering (Degree in AE, CD, EE, EM, ME, or Physics with related experience)

Analyze, develop, perform, and evaluate tests for all programs requiring the engineering test laboratories. Establish for each laboratory test; schedules, objectives, methods, data acquisition techniques and presentation, test facilities and equipment required, and cost estimates for the test performed. Design or innovate new techniques and equipment when required to meet new or unusual test requirements. Engineering test laboratories includes stress analysis, structural testing, systems test engineering, heat transfer, flight test, wind tunnel, propulsion systems, propellant research, instrumentation vibration, acoustics, and vacuum chambers.

Training Equipment Engineering (Degree in AE, EE, ME, or CE)

Analyze, design, and develop electronic and mechanical training equipment, which is system oriented and portable for field use. Provide technical assistance during fabrication and testing of the equipment. Prepare operating and functional test manuals for the equipment. May help gather data for training proposals, prepare engineering work statements, and coordinate with other technical groups within the company.

Training Simulators (Degree in CS, Math, EE, or Physics)

Design and develop man-in-the-loop simulators for training and research. The applied technologies include computer image generation, courseware authoring systems, real-time application software, high-fidelity simulation and firmware, as well as the implementation of components of systems on digital computers. Individual assignments may encompass one or more of these technologies.

Weight Engineering (Degree in AE, CE, EE, EM, or ME)

Perform engineering analysis and participate in every aspect of the design to ensure the minimum possible weight consistent with established cost, reliability, and performance objectives. Conduct weight and balance trade studies on new vehicles under consideration in preliminary design groups; establish design target weights, and maintain effective weight control during the designing and building phases to assure weight compliance; establish vehicle balance and loadability characteristics; investigate problems in mass distribution, moment of inertia, mass properties for simulators, control studies, and other mass balance problems in new vehicle design, production, and flight testing operations. Involves continual interaction with other engineering disciplines in every facet of aircraft, missile, and spacecraft design, such as structures, power plant, electronics, aerodynamics, systems, and components.

Wind Tunnel Engineering—Instrumentation (Degree in EE)

Provide and maintain data acquisition and data-handling capability for the wind tunnel testing activities. Research and develop new instrumentation methods for measuring aerodynamic forces, pressures, heat transfer, and dynamic loads.

Wind Tunnel Engineering—Testing (Degree in AE or ME)

Provide wind tunnel support services for all company organizations involving the planning, conducting, and reporting of aircraft, missile, and space vehicle model tests in low speed, transonic, supersonic, and hypervelocity tunnels. Manage the data flow process throughout all test programs.

Career Profile

Engineering

Name:		Peter Clark
Career Field:		Engineering
Position Held:		Airborne Engineering Support Department Head Electrospace Systems Inc., Richardson, Texas
Education:	**1969**	Masters in Electrical Engineering U.S. Air Force Institute of Technology
	1964	Bachelors in Electrical Engineering Oklahoma University, Stillwater, Oklahoma

Experience:

1983 - Present: Electrospace Systems, Inc., Richardson, TX
Department Head/Staff Engineer

Company performs integration of avionic systems on various commercial and military aircraft. Systems installed include navigation, communication, instrument, and special mission equipment. Job responsibilities involve program management duties as well as overseeing day-to-day engineering operations.

1979 - 1983: Kentron International, Inc., Dallas, Texas
Corporate Director, National Operations

For contracts with the Western United States, was responsible for profit/loss, proper staffing, corporate interface with customers, corporate representation at negotiations, contract performance, development of strategy for re-win and labor relations. Had eleven contracts with NASA, USAF, and USN. Reported to Executive Vice President.

1974 - 1979: U.S. Air Force
Chief, Operating Location, Electrical Engineer and Project Officer,
U.S. Army Dugway Proving Ground, Utah

Responsible for securing all support for USAF tests of drone aircraft at Army proving ground. Support ranged from telemetry to communications to recovery of crashed vehicles. Also served as Project Officer for $4M contractor operations and maintenance effort. Wrote SOW portion of contract. Leader of source selection committee for Utah portion of contract.

Responsible for design of VHF non-tactical communications system for the Utah range. Served as Project Officer for testing airborne ELINT package for HF through SHF. Designed backup HF communications system for long-range tests from Edwards AFB into Utah.

1972 - 1974: U.S. Air Force

Chief of Special Projects at Edwards AFB, California and then transferred with organization to Hill AFB, Utah.

Responsible for development of electronic equipment/systems to correct deficiencies on drone aircraft. Responsible for flight test range electronic equipment installation/upgrade for move of the range from Edwards to Utah. Designed standardized range safety kit for Teledyne-Ryan-34 series of Remotely Piloted Vehicles.

1969 - 1972: U.S. Air Force

Project Officer, Data Reduction Center and Tactical Operation Room

Responsible for hardware portion of satellite ground station, including two IBM 360-75 computers with special pre-processor. Assigned one year in Los Angeles and two years in Australia. Duties were contractor selection of computer system, tape recorder system with bit error checking, software acceptance testing, hardware acceptance testing, establishment of precision measurement equipment laboratory, evaluation of contractor EMP vulnerability study and served as Chief of Quality Assurance.

1964 - 1967: U.S. Air Force

Airborne Instrumentation Development Project Officer, Edwards AFB, California

Responsible for development of non-standard aircraft instrumentation used in flight test of aircraft. Projects worked on were XB-70, XV-5A, XC-142, F-111, and C-130A. Responsible for interfacing instrumentation to telemetry link to ground.

1957 - 1962: U.S. Air Force

Early Warning Radar Operator

Career Synopsis

My career began with a tour as an enlisted radar operator in the United States Air Force. The USAF had a program through which selected enlisted men could obtain a college degree and then a commission. After completion of this program and commissioning, I was stationed at Edwards AFB, California, where the assignment required significant interaction with contractors and their associated contractual documents. Throughout my career, this was the general requirement for each assignment, although some engineering was accomplished. This need for contractual knowledge proved extremely beneficial in civilian life because so much of higher level civilian assignments have active involvement in contractual matters.

After the assignment at Edwards AFB, the USAF assigned me to a program through which a master's degree in electrical engineering was obtained. For the remainder of my career, I had various assignments, all of which were in the engineering field, but where most of the emphasis was on contractor support to the USAF. These types of assignments prepare an individual very well for civilian work with companies that are in the defense business. After twenty-two years of service, I retired from the USAF at the rank of Major.

At that time, I obtained a position with a service company that specialized in supplying complete facility operations for the government. Some examples are operation of the U.S. Army Yuma Proving Ground data reduction center and operation of the Hill AFB test range. The company did not supply personnel in the mode of the temporary services. When they hired me, they were looking for personnel with precisely the type of experience discussed above, and they gave no thought to training any of their personnel to promote from within the company.

After four years, I accepted a position with Electrospace Systems, Inc., an aircraft modification company, once again, using the experience gained from service in the USAF as a basis for employment. The four years with the service company was further beneficial, as it broadened my experience in the government contracting fields. This company has progressed from small to medium size during the last five years. Their interest in employees is for well-rounded people and they provide the opportunity for engineers to acquire training in all areas through "on-the-job" training. They take promising engineers and assign them as Program Managers (usually having them perform as lead engineer, as well). In this assignment, the people are exposed to all areas of their project and make the major decisions affecting program direction and success.

Recommendations to Job Hunters:

There are many interesting and challenging engineering jobs available throughout this country to qualified individuals. Typically, most engineering positions require a Bachelor's degree and technician type positions require an Associate's degree or equivalent experience.

One problem facing many college graduates interested in employment is experience. Many small engineering companies want to hire experienced people. They do not have the time to train new engineers on job responsibilities. Larger organizations, on the other hand, will typically hire and train the applicant. They can afford the six- to eight-month learning period when the employee is fairly unproductive.

The military is an excellent training ground for people interested in aviation/aerospace type jobs. Not only will they send the applicant to a technical training school and spend thousands of dollars training you, and pay you at the same time, but they provide hands-on experience that is one of the most valuable assets a potential job hunter can have. It is possible to go this route through the civilian sector, but it will then become an out-of-pocket expenditure. Civilian education is not cheap, but it is available to those who can afford it. The military also offers VA college benefits to those who satisfy their military tour.

This same situation is also true in upper-level engineering positions. Civilian companies generally do not train their personnel but rather seek experienced personnel for the middle management positions, and what they seek are those personnel who have well-rounded experience. This "well rounding" is not as common in civilian industry where the trend is to essentially keep personnel in the area where that person works. This is particularly true in the bigger companies that force very narrow specialization upon their personnel. Rarely does an engineer have the contractual experience in a civilian company to progress past lower levels of management. The ones who do have it obtained the necessary training on their own, usually through an MBA program.

Overspecialization is a potential problem in any career field, but it's especially true in engineering. This problem is especially applicable to the recent college graduate with no real experience. Large organizations typically hire and train the individual to do a specific job. This can lead to a very specialized, "non-rounded employee." The long-term effect could be a dead-end job with little or no promotion possibility.

The best deterrent to this situation is to be aware of it and plan your career goals accordingly. Well-roundedness is a very important asset, and it becomes more and more critical as your career progresses.

Engineering, in general, is a highly creative and very challenging career field. People who work in engineering are in a very visible position for promotion if qualified. Engineers can usually enjoy the benefit of upward mobility into management.

Also, a benefit is that many employers provide financial assistance to employees who are interested in furthering their education. They do this as an incentive to keep their valued employees and to help create career paths for motivated professionals within the organization.

Depending on the position, some engineering positions such as field service engineers or technical support representatives require extensive travel and field work. This could be a benefit for the single professional who enjoys traveling or a drawback to the employee who has a family and does not want to travel. These are all things that should be considered when choosing a career field.

Education, training, experience, and motivation all are important aspects to anyone who is pursuing a career as an engineer. If properly planned and executed, an engineering profession can be a intellectually and financially rewarding way of life.

AEROSPACE INDUSTRY COMPANIES

Table 3-1 lists current addresses of a large cross-section of employers in the aerospace industry.

Table 3-1. AEROSPACE INDUSTRY COMPANIES

AERO MODIFICATIONS
2201 Scott Ave.
Fort Worth, TX 76103
(817)535-1936

AEROJET GENERAL
10300 N. Torrey Pines Rd.
La Jolla, CA 92037
(619)455-8668

AERONCA, INC.
Aerospace Division
1712 Germantown Rd.
Middletown, OH 45042

AEROSPACIAL AIRCRAFT CORP.
1100 Vermont Ave., N. W.
Washington, D.C. 20005
(202)331-8980
(Note: U.S. Subsidiary-ATR 42)

AEROSPATIALE HELICOPTER CORP.
2701 Forum Dr.
Grand Prairie, TX 75053-4005
(214)641-0000
(Note: U.S. Subsidiary)

AGUSTA AVIATION CORP.
Norcom & Red Lion Rds.
Philadelphia, PA 19154
(215)245-8860
(Note: U.S. Subsidiary)

AHRENS AIRCRAFT CORP.
2800 Teal Club Rd.
Oxnard, CA 93030
(805)985-2000

AIR TRACTOR, INC.
Box 485
Olney, TX 76374
(817)564-5616

ALLIED SIGNAL AEROSPACE CO.
2525 W. 190th St.
Torrance, CA 90509
(213)532-7151

ALLISON GAS TURBINE DIVISION
General Motors Corp.
2001 Tibbs St.
Indianapolis, IN 46206
(317)230-2000

AMERICAN AIRLINES, INC.
Maintenance, Engineering, Data Processing & Communications Center
3800 N. Mingo Rd.
Tulsa, OK 74158-2809

AMERICAN AVIATION INDUSTRIES (AAI)
16700 Roscoe Blvd.
Van Nuys, CA 91406
(818)786-1921

AMETEK, INC.
21st Floor, 410 Park Ave.
New York, NY 10022
(212)935-8640

ANDREW CORP.
10500 W. 153rd St.
Orland Park, IL 60462

ANSER
Suite 800
1215 Jefferson Davis Hwy.
Arlington, VA 22202
(703)685-1000
1-800-368-4173

ARCTIC AIRCRAFT CO.
Box 6-141
Anchorage, AK 99502
(907)243-1580

AVCO CORP.
1275 King St.
Greenwich, CT 06830
(203)552-1800

AVCO (AEROSTRUCTURE DIVISION)
Box 210
Nashville, TN 37202
(615)361-2000

AVCO LYCOMING
Stratford Division
550 S. Main St.
Stratford, CT 06497
(203)378-8211

AVTEK CORP.
4680 Calle Carga
Camarillo, CA 93010
(805)482-2700

AYRES CORP.
Box 3090
Municipal Airport
Albany, GA 31708
(912)883-1440

BALL AEROSPACE SYSTEMS DIVISION
Box 1062
Boulder, CO 80306
(303)939-4000

BDM INTERNATIONAL INC.
7915 Jones Branch Dr.
McLean, VA 22102

BEECH AIRCRAFT CORP.
Box 85
Wichita, KS 67201-0085
(316)681-7111

BELL AEROSPACE TEXTRON
Box 1
Buffalo, NY 14240
(716)297-1000

BELLANCA AIRCRAFT CORP.
Box 69
Municipal Airport
Alexandra, MN 56308
(612)762-1501

BELLANCA AIRCRAFT ENGINEERING, INC.
Galena, MD 21635
(301)648-5172

BELL HELICOPTER, TEXTRON INC.
Box 482
Fort Worth, TX 76101
(817)280-2011

THE BENDIX CORP.
Bendix Center
Southfield, MI 48076
(313)352-5000

BENDIX/KING AIR TRANSPORT
Avionics Division
P.O. Box 9327, 2100 NW 62nd St.
Fort Lauderdale, FL 33310
(305)928-2100

BENDIX/KING GENERAL AVIATION
Avionics Division
400 N. Rogers Rd.
Olathe, KS 66062

BOEING AEROSYSTEMS INTL.
Box 3707
Renton, WA 98124
(206)237-3289

THE BOEING CO.
Box 3707
Seattle, WA 98124-2207
(206)655-2121

BOEING MILITARY AIRPLANE CO.
P.O. Box 7730
Wichita, KS 67277-7730

BOEING SUPPORT SERVICE INTL.
Box 3707
Renton, WA 98128
(206)237-3284

BOEING TECHNOLOGY SERVICES INTL.
Box 3707
Renton, WA 98128
(206)237-2722

BOEING VERTOL CO.
P.O Box 16858
Boeing Center
Philadelphia, PA 19142
(215)522-2121

BRITISH AEROSPACE, INC.
Dulles International Airport
Washington, D.C. 20041
(703)435-9100
(Note: U.S. Subsidiary)

BURROUGHS CORP.
Burroughs Place
Detroit, MI 48232
(313)972-7000

CAMAIR AIRCRAFT CORP.
Box 600
Remsenburg, NY 11960

CESSNA AIRCRAFT CO.
Box 1521
Wichita, KS 67201
(316)685-9111

CONRAC CORP.
Three Landmark Square
Stamford, CT 06901
(203)348-2100

CURTIS WRIGHT CORP.
1200 Wall St. W
Lyndhurst, NJ 07071
(210)896-8400

DE HAVILLAND AIRCRAFT OF CANADA, LTD.
Garratt Blvd.
Downsview, Ontario, Canada
M3K 1Y5
(416)633-7310

DORNIER AVIATION (NORTH AMERICA), INC.
1213 Jefferson Davis Hwy. Suite 1001
Arlington, VA 22202-4304
(703)769-7228
(Note: U.S. Subsidiary)

DOUGLAS AIRCRAFT CO.
3855 Lakewood Blvd.
Long Beach, CA 90846
(213)593-5511

EATON CORP.
One Huntington Quadrangle
Melville, NY 11747
(516)293-8900

ELECTROSPACE SYSTEMS INC.
P.O. Box 831359
Richardson, TX 75038-1359
(214)470-2000
(Subsidiary of the Chrysler Corp.)

EMBRAER AIRCRAFT CORP.
276 SW 34th St.
Fort Lauderdale, FL 33315
(305)524-5755
(Note: U.S. Subsidiary)

EMERSON ELECTRIC CO.
Government and Defense Group
8100 W. Florissant Ave.
St. Louis, MO 63136
(314)553-2000

ENSTROM HELICOPTER CORP.
P.O. Box 277
Menominee, MI 49858
(906)863-9971

E-SYSTEMS
Greenville Division
Box 1056
Greenville, TX 75401
(214)455-3450

FAIRCHILD AEROSPACE PRODUCTS DIVISION
8463 Higuera St.
Culver City, CA 90230
ATTN: T.L. Collins
Employee Relations
(213)202-8200

FAIRCHILD AIRCRAFT CORP.
P.O. Box 790490
San Antonio, TX 78279
ATTN: P.T. Granato
Human Resources Director
(512)824-9421

FAIRCHILD COMMUNICATIONS & ELECTRONICS CO.
Sherman Fairchild Technology Center
20301 Century Blvd.
Germantown, MD 20874-1182
ATTN: T.L. Bedocs
Human Resources & Administration
(301)428-6000

FAIRCHILD CONTROL SYSTEMS CO.
1800 Rosecrans Ave.
Manhattan Beach, CA 90266-3797
ATTN: W.H. Adams
Human Resources & Administration
(213)643-9222

FAIRCHILD REPUBLIC CO.
Farmingdale, NY 11735
(516)531-0105

FAIRCHILD SPACE CO.
Sherman Fairchild Technology Center
20301 Century Blvd.
Germantown, MD 20874-6000
ATTN: T.L. Bedocs
Human Resources & Administration
(301)428-6000

FALCON JET CORP.
Teterboro Airport
Teterboro, NJ 07608
(201)288-5300

FOKKER AIRCRAFT USA, INC.
1199 N. Fairfax St.
Suite 500
Alexandra, VA 22314
(703)838-0100
(Note: U.S. Subsidiary)

FORD AEROSPACE & COMMUNICATIONS CORP.
Box 43342
300 Renaissance Center, 20th Floor
Detroit, MI 48243
(313)446-7660

FORD AEROSPACE & COMMUNICATIONS
Western Development Labs.
3939 Fabian Way
Palo Alto, CA 94303-9981
(415)852-4965

FORD AEROSPACE & COMMUNICATIONS CORP.
2715 Nevada Ave.
Norfolk, VA 23513-4412

THE GARRETT CORP.
Box 92248
9851 Sepulveda Blvd.
Los Angeles, CA 90009
(213)776-1010

GATES LEARJET CORP.
Box 7707
Mid-Continent Airport
Wichita, KS 67277
(316)946-2000

GENERAL AIRCRAFT CORP.
Hanscom Field
Bedford, MA 01730
(617)274-9130

GENERAL DYNAMICS CORP.
Pierre Laclede Center
St. Louis, MO 63105
(314)889-8200

GENERAL ELECTRIC CO.
3135 Easton Turnpike
Fairfield, CT 06431
(203)373-2211

GOODYEAR AEROSPACE CORP.
1210 Massillon Rd.
Akron, OH 44315
(216)796-2121

GREAT LAKES AIRCRAFT CO.
Offices & Production Plant:
Drawer A
Eastman, GA 31023
(912)374-5535

GRIMES DIVISION—MIDLAND ROSS CORP.
550 State Route 55
Urbana, OH 43078

GRUMMAN CORP.
1111 Steward Ave.
Bethpage, NY 11714-3580
(516)575-0574

GULFSTREAM AMERICAN CORP.
Box 2206
Savannah International Airport
Savannah, GA 31402
(912)964-3000

HAMILTON STANDARD, DIV. OF UTC
Bradley Field Rd.
Windsor Locks, CT 06096
(203)623-1621

HARRIS CORP.
1025 W. NASA Blvd.
Melbourne, FL 32919
(305)727-9100

HARRIS CORP.
Government Communications
Systems Group
P.O. Box 92000
Melbourne, FL 32902

HONEYWELL INC.
Honeywell Plaza
Minneapolis, MN 55408
(612)870-5200

HUGHES AIRCRAFT CO.
P.O. Box 45066
7200 Hughes Terrace
Los Angeles, CA 90045-0066
(213)568-7200

INTERNATIONAL BUSINESS MACHINES CORP.
Old Orchard Rd.
Armonk, NY 10504
(914)765-1900

ITT CORP.
320 Park Ave.
New York, NY 10022
(212)752-6000

ITT GENERAL CONTROLS/ AEROSPACE PRODUCTS
1200 South Flower Ave.
Burbank, CA 91502
(818)953-2073

KAMAN CORP.
Blue Hills Ave.
Bloomfield, CT 06002
(203)243-8311

KAMAN SCIENCES CORP.
1500 Garden of the Gods Rd.
P.O. Box 7463
Colorado Springs, CO 80933-7463
(303)599-1500

KING RADIO CORP.
400 North Rogers Rd.
Olathe, KS 66062
(913)782-0400

KOLLSMAN INSTRUMENT CO.
Div-Sun Chemical Corp.
Daniel Webster Hwy. S.
Merrimack, NH 03054
(603)889-2500

LEAR SIEGLER INC.
P.O. Box 2158
2850 Ocean Park Blvd.
Santa Monica, CA 90406
(213)452-5444

LINK SIMULATOR DIVISION
Corporate Dr.
Binghamton, NY 13902
(607)772-3011

LITTON COMPUTER SERVICES
1300 Villa St.
P.O. Box 7133
Mountain View, CA 94039-7113
(415)966-1771

LITTON INDUSTRIES
360 N. Crescent Dr.
Beverly Hills, CA 90210-4867
(213)859-5000

LOCKHEED CORP.
Corporate Offices
4500 Park Granada Blvd.
Calabasas, CA 91399-0510
(818)712-2416

Lockheed Aeronautical Systems Group

LOCKHEED—CALIFORNIA CO.
Dept. 90-10
Burbank, CA 91520
(818)847-4723

LOCKHEED—GEORGIA CO.
86 South Cobb Dr.
Marietta, GA 30063
(404)424-4411

MURDOCK ENGINEERING CO.
P.O. Box 152278
Irving, TX 75015
(214)790-1122

LOCKHEED AIRCRAFT SERVICE CO.
Ontario International Airport
P.O. Box 33
Ontario, CA 91762-8033
(714)395-2411

LOCKHEED AEROMOD CENTER
1044 Terminal Rd.
Greenville, SC 29605
(803)299-3350

LOCKHEED SUPPORT SYSTEMS, INC.
1600 E. Pioneer Parkway
Arlington, TX 76010
(800)433-5339

LOCKHEED AIR TERMINAL
P.O. Box 7229
Burbank, CA 91510
(818)503-1592

LOCKHEED ADVANCED AERONAUTICS CO.
Dept. 60-30
Burbank, CA 91520
(805)257-5775

LOCKHEED—ARABIA
P.O. Box 33
Dept. 1-782
Ontario, CA 91762-8033
(714)988-2531

Lockheed Information Systems Group

CADAM, INC.
1935 North Buena Vista
Burbank, CA 91504
(818)841-9470

DATACOM SYSTEMS CORP.
Glenpointe Center East
300 Frank W. Burr Blvd.
Teaneck, NJ 07666
(201)692-2900

DIALOG INFORMATION SERVICES
3460 Hillview Ave.
Palo Alto, CA 94304
(415)858-2700

LOCKHEED DATAPLAN, INC.
90 Albright Way
Los Gatos, CA 95030

Lockheed Missiles, Space, and Electronics Systems Group

LOCKHEED MISSILES & SPACE CO.
P.O. Box 3504
(408)743-2200
and
P.O. Box 17100
Austin, TX 78760
(512)448-5555

LOCKHEED SPACE OPERATIONS CO.
P.O. Box 2807
Titusville, FL 32780
(305)383-2200
and
1100 W. Laurel Ave.
Lompoc, CA 93436
(805)735-7711

LOCKHEED ENGINEERING & MANAGEMENT SERVICES CO.
2400 NASA Rd. 1
P.O. Box 58561
Houston, TX 77058
(713)333-5411

LOCKHEED ELECTRONICS CO.
1501 U.S. Hwy. 22
Plainfield, NJ 07061
(201)757-1600

LSI AVIONIC SYSTEMS CORP.
7-11 Vreeland Rd.
Florham Park, NJ 07932-0760
(201)822-1300

THE LTV CORP.
LTV Tower
Box 655003, 2100 Ross Ave.
Dallas, TX 75265-5003
(214)979-7711

LTV AEROSPACE & DEFENSE CO.
Vought Aero Products Division
P.O. Box 225907
Dallas, TX 75265
ATTN: Professional Placement
(214)266-2011

LTV AEROSPACE & DEFENSE CO.
Vought Missiles & Advanced Programs Div.
P.O. Box 650003
Dallas, TX 75265-0003
ATTN: Professional Placement
(214)266-2011

LTV AEROSPACE & DEFENSE CO.
Vought Missiles & Advanced Programs Div.
Camden Operations
P.O. Box 1015
Camden, AR 71701

MAGNAVOX GOVERNMENT & INDUSTRIAL ELECTRONICS CO.
Hangar 19
Ft. Wayne, IN 46808
(219)429-6000

MARTIN MARIETTA AEROSPACE
6801 Rockledge Dr.
Bethesda, MD 20817
(301)897-6000

MARTIN MARIETTA AEROSPACE
Baltimore Division
103 Chesapeake Park Plaza
Baltimore, MD 21220
(301)338-5000

MARTIN MARIETTA AEROSPACE
Denver Division
Box 179
Denver, CO 80201
(303)977-3000

MARTIN MARIETTA AEROSPACE
Orlando Division
Box 5837
Orlando, FL 32855
(305)352-2000

MBB HELICOPTER CORP.
900 Airport Rd.
P.O. Box 2349
West Chester, PA 19380
(215)431-4150

MCDONNELL DOUGLAS CORP.
Box 516
St. Louis, MO 63166
(314)232-0232

MCDONNELL DOUGLAS HELICOPTORS
5000 E. McDowell
Mesa, AZ 85205
(602)891-2301

MOONEY AIRCRAFT CORP.
Sub-Republic Steel Corp.
Louis Schreiner Field
Box 72
Kerrville, TX 78029-0072
(512)896-6000

MOTOROLA INC.
1303 East Algonquin Rd.
Schaumburg, IL 60196
(312)397-5000

NORDEN SYSTEMS
Norden Place
Norwalk, CT 06856
(203)852-5000

NORTH AMERICAN AIRCRAFT
(Div. of Rockwell)
Headquarters
Box 92098
Los Angeles, CA 90029
(213)647-1000

NORTHROP CORP.
1840 Century Park East
Century City
Los Angeles, CA 90067-2199
(213)332-2000

NORTHROP CORP.
Aircraft Division
1 Northrop Ave.
Hawthorne, CA 90250

PAN AM WORLD SERVICES INC.
Aerospace Division
Eastern Test Range Project
P.O. Box 4608
Patrick AFB, FL 32925

PARKER BERTEA AEROSPACE GROUP
ATTN: Personnel Dept.
18321 Jamboree Blvd.
Irvine, CA 92715
(714)833-3000

PARKER HANNIFIN CORP.
Corp. Hdqtrs:
17325 Euclid Ave.
Cleveland, OH 44112
(216)531-3000

PARTENAVIA OF AMERICA
1235 Jefferson Davis Hwy.
Arlington, VA 22202
(703)243-1700
(Note: U.S. distributor)

PERKIN-ELMER, COMPUTER SYSTEMS DIV.
Two Crescent Pl.
Oceanport, NJ 07757
(201)229-6800

PIPER AIRCRAFT CORP.
P.O. Box 1328, 2926 Piper Dr.
Vero Beach, FL 32960
(305) 567-4361

PITTS AEROBATICS
Box 542
Afton, WY 83110
(305)247-5423

PRATT & WHITNEY AIRCRAFT CORP.
400 Main St.
East Hartford, CT 06108
(203)565-4321

RADIAN CORP.
8501 Mopac Blvd.
Austin, TX 78759

RAND CORP.
1700 Main St.
Santa Monica, CA 90406

RAYTHEON CO.
141 Spring St.
Lexington, MA 92173
(617)862-6600

RAYTHEON MISSILE SYSTEMS DIVISION
Hartwell Rd.
Bedford, MA 01730
(617)274-7100

RCA
30 Rockefeller Plaza
New York, NY 10020
(212)598-5900

ROBINSON HELICOPTER CO., INC.
24747 Crenshaw Blvd.
Torrance, CA 90505
(213)539-0508

Rockwell International Corp.

HEADQUARTERS
600 Grant St.
Pittsburgh, PA 15219

HEADQUARTERS (AEROSPACE DIV.)
2230 E. Imperial Hwy.
El Segundo, CA 90245
(213)647-5000

Rockwell Space Divisions

ROCKWELL HANFORD OPERATIONS
P.O. Box 800
Richland, WA 90009

ROCKWELL INTERNATIONAL
Space Transportation & Sys. Div.
College Relations Office

12214 Lakewood Blvd.
Downey, CA 90241

1840 NASA Rd. I
Houston, TX 77058

ROCKWELL INTERNATIONAL
Satellite Systems Division
College Relations Office
2600 Westminister Ave.
Seal Beach, CA 90740

ROCKWELL INTERNATIONAL
Rocketdyne Division
College Relations Office
6633 Canoga Ave.
Canoga Park, CA 91304

ROCKWELL INTERNATIONAL
Energy Systems
College Relations Office

Rocky Flats Plant
P.O. Box 464
Golden, CO 80401

Rockwell Aircraft Divisions

ROCKWELL INTERNATIONAL
North American Aircraft Operations
College Relations Office

P.O. Box 92098
Los Angeles, CA 90009

4300 East Fifth Ave.
P.O. Box 1259
Columbus, OH 43216

2825 E. Ave. P
Palmdale, CA 93550

ROCKWELL INTERNATIONAL
Tulsa Division
College Relations Office
P.O. Box 51308
2000 N. Memorial Dr.
Tulsa, OK 74151

Rockwell Commercial Electronics Divisions

ROCKWELL INTERNATIONAL
Avionics Group
College Relations Office
MS 105-180
400 Collins Rd. N.E.
Cedar Rapids, IA 52948

ROCKWELL INTERNATIONAL
Collins Transmission Sys. Div.
College Relations Office
P.O. Box 10462
M.S. 401-152
Dallas, TX 75207

ROCKWELL INTERNATIONAL
Switching Sys. Div.
College Relations Office
1431 Opus Place
Downers Grove, IL 60515

ROCKWELL INTERNATIONAL
Wescom Telephone Products Div.
College Relations Office
8245 Lemont Rd.
Downers Grove, IL 60516

ROCKWELL INTERNATIONAL
Semiconductor Products Div.
College Relations Office
4311 Jamboree Rd.
Newport Beach, CA 92660

Rockwell Defense Electronics

ROCKWELL INTERNATIONAL
Missiles Systems Div.
College Relations Office
1800 Satellite Blvd.
Duluth, GA 30136

ROCKWELL INTERNATIONAL
Autonetics Divisions
College Relations Office
3370 Miraloma Ave.
Anaheim, CA 92803

ROCKWELL INTERNATIONAL
Microelectronics Research & Development Center
College Relations Office
3370 Miraloma Ave.
Anaheim, CA 92803

ROCKWELL INTERNATIONAL
High Frequency Communications Div.
College Relations Office
3200 E. Renner Rd.
Richardson, TX 75081

Rockwell International
Advanced Communications &
Countermeasures Division
2901 W. MacArthur Blvd.
P.O. Box 11963
Santa Ana, CA 92711

Rockwell International
Advanced Communications &
Countermeasures Division
College Relations Office
855 35th St. N.E.
Cedar Rapids, IA 52498

Saab Aircraft of America, Inc.
P.O. Box 17188
Dulles International Airport
Washington, D.C. 20041
(703)478-9720
(Note: U.S. Subsidiary)

Short Brothers USA, Inc.
2011 Crystal Dr.
Suite 713
Arlington, VA 22202
(703)769-5555
(Note: U.S. Subsidiary)

The Singer Co.
Aerospace & Marine Systems
8 Stamford Forum
Stamford, CT 06904
(203)356-4200

Singer Kearfott
Kearfott Guidance/Navigation Div.
1225 McBride Ave.
Little Falls, NJ 07424
(201)785-6000

Sikorsky Aircraft
6900 Main St.
Stratford, CT 06601-1381
(203)386-4000

Simmonds Precision Products Inc.
150 White Plains Rd.
Tarrytown, NY 10591
(914)631-7500

Sperry Commercial Flight
Systems Division
P.O. Box 29000
Phoenix, AX 85038
(602)863-8000

Sperry Corp.
1290 Ave. of the Americas
New York, NY 10104
(212)484-4444

Sperry Flight Systems
Box 21111
Phoenix, AZ 85036
(602)869-2311

Sperry Corp—Defense
Products Group
3333 Pilot Knob Rd.
Eagan, MN 55164

Sperry Corp
Systems Mgt. Group
12010 Sunrise Valley Dr.
Reston, VA 22021

Taylorcraft Aviation Corp.
Box 947, 820 E. Bald Eagle St.
Lock Haven, PA 17745
(717)748-6712

Teledyne CAE
1330 Laskey Rd.
Toledo, OH 43612
(419)470-3000

Teledyne Ryan Aeronautical
2701 Harbor Dr.
San Diego, CA 92138
(619)291-7311

Texas Instruments Inc.
13500 N. Central Expressway
Dallas, TX 75265
(214)995-2011

TRW Electronics & Defense
One Space Park
Redondo Beach, CA 90278
(213)535-1235

TRW INC.
23555 Euclid Ave.
Cleveland, OH 44117
(216)383-3178

UNITED TECHNOLOGIES
Government Products Div.
West Palm Beach, FL 33402
(305)840-2000

UNITED TECHNOLOGIES CORP.
United Technologies Bldg.
Hartford, CT 06101
(203)728-7000

VERTICAL LIFT TECHNOLOGIES, INC.
1830 W. Drake Dr.
Tempe, AZ 85283
(602)345-7855

VOUGHT AERO PRODUCTS DIVISION
(See LTV Aerospace)

VOUGHT MISSILES AND ADVANCED PROGRAMS DIVISION
(See LTV Aerospace)

WESTERN GEAR CORP.
P.O. Box 22674
Long Beach, CA 90801
(213)595-9556

WESTINGHOUSE ELECTRIC CORP.
Westinghouse Bldg.
Gateway Center
Pittsburgh, PA 15222
(412)255-3800

WESTON INSTRUMENTS
Sangamo-Weston, Inc.
614 Frelinghuysen Ave.
Newark, NJ 07114
(201)242-2600

4
Flight Operations

Boeing 737

The airline industry in the United States presently consists of approximately 250 commercial operators flying more than 4,500 aircraft and employing more than 355,000 people. Major, national, and regional airlines are known as *commercial air carriers* and conduct scheduled services on specific routes plus nonscheduled, or charter, operations.

The term *commuter air carriers*, on the other hand, refers to carriers that operate aircraft with a maximum of 60 seats, make at least five round trips between two points weekly, or carry mail.

The Future Aviation Professionals of America (FAPA), an Atlanta-based information service specializing in employment data for pilots, mechanics, and flight attendants, forecasts that in the next decade, there will be 52,000 pilot job openings for large jets with an additional 10,000 openings in the regional carriers.

The new record was set in airline hiring in 1985 when a record of 7,850 pilots were hired by existing, or start-up, airlines. The previous record was set in 1984 when 5,700 pilots were hired. The record previous to 1984 was set in 1966 when slightly over 4,500 pilots joined the ranks. According to FAPA, the hiring trend will continue into the next decade. It is also forecasted that many more jobs will be created by corporate, business, government, and general aviation organizations.

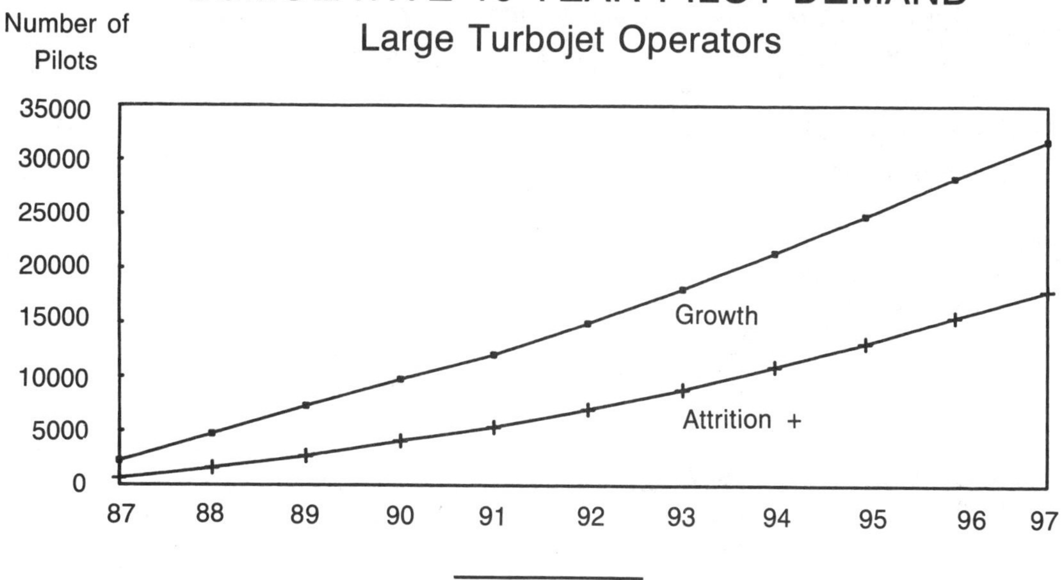

FAPA, Atlanta, GA 1-800-JET-JOBS

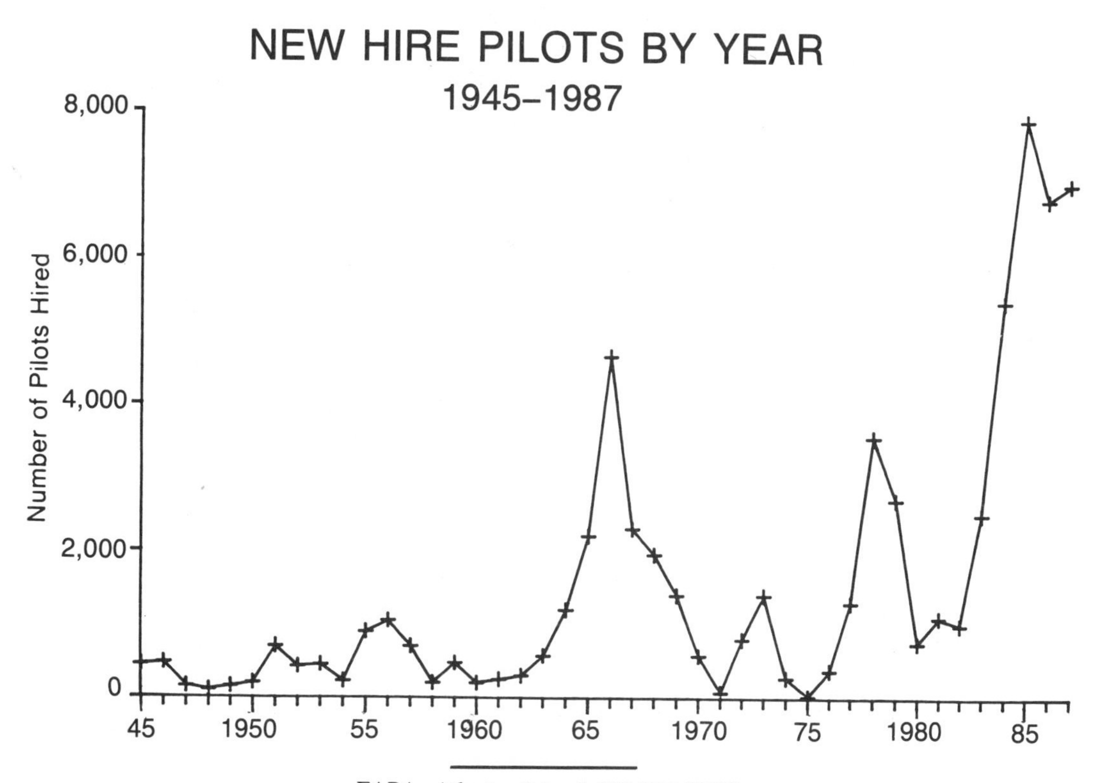

FAPA, Atlanta, GA 1-800-JET-JOBS

The aviation industry is very sensitive to economic trends. Operating profits and losses closely parallel periods of economic growth or recession. It is often noted that the performance of the airlines is directly linked to the overall U.S. economic climate.

SO YOU WANT TO SIT IN THE COCKPIT?

Aircraft pilots understandably have a strong attachment to their occupation because it requires a substantial amount of time, effort, and training. According to the U.S. Department of Labor, the employment of pilots is expected to increase faster than the average for all occupations through the mid 1990s.

With the expected growth in airline passenger and cargo traffic, the demand for more aircraft, pilots, flight instructors, and other flying personnel is on the rise. Businesses are expected to operate more aircraft and hire more pilots and crew members to fly passengers and cargo to the increased number of locations that the scheduled airlines do not service.

After WWII and the Korean War, many military pilots joined the ranks of commercial pilots now serving the airline industry. With many of these pilots retiring, the need for qualified pilots has increased substantially.

The Airline Pilots Association (ALPA) predicts that 7,000 pilots now flying will reach the mandatory retirement age by 1995 when annual retirees will peak at 2,200. Data also shows that 90 percent of these will be pilots from the major airlines. This trend will provide upward mobility for many younger pilots. They estimate that in 1989, the number of pilots retiring will exceed 1,000, in 1991 it will reach 1,500, in 1993 it will rise to 2,000, and by 1995, 2,200 pilots will be forced to leave their posts.

In the past, the military has been the traditional training grounds for airline pilots. In recent years the military has been offering incentives such as better pay, benefits, and working conditions to encourage reenlistment.

It is interesting to note that in late 1985, more civilian pilots were hired into the major airlines than military pilots, but in late 1986, the number of military pilots hired increased dramatically.

Of the military pilots hired in 1987, the Air Force contributes 55 percent, the Navy 28 percent, and the Army, Marines, and Coast Guard provide the remaining 17 percent.

Many pilots in the general aviation market start off as flight instructors, earning some money while they build up their flying hours. As these pilots become more experienced, they tend to move on to more advanced flying jobs such as air charter, air taxi, and business flying operations. On a smaller scale, some pilots provide services such as crop dusting, banner towing, inspecting pipelines, and conducting sight-seeing trips.

With this valuable training and experience, many of these pilots then move on to higher paying pilot positions with the major and national airlines. This progression creates vacancies at the lower ranks for new pilots.

To become a commercial pilot, the applicant can either go the military route and enlist for the required tour, or he/she can obtain the FAA ratings necessary in the civilian sector.

AVIATION BACKGROUND
Major Airline New Pilot Hire 1977-88

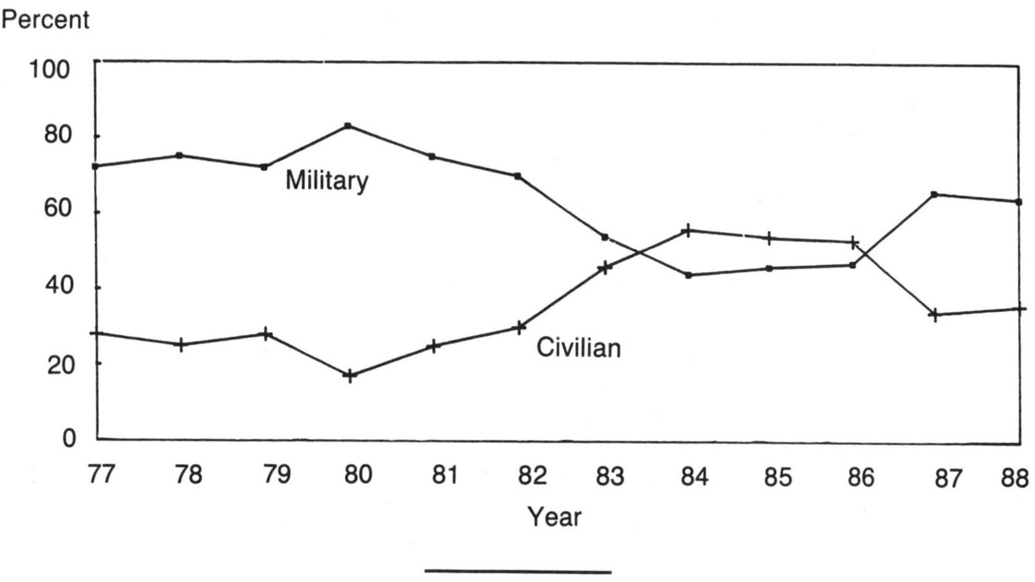

FAPA, Atlanta, GA 1-800-JET-JOBS

The military training offers many financial advantages. Not only does the applicant receive the finest formal flight training, but he/she gets paid to do so. Earning your wings in the civilian sector can be an expensive proposition, but there are many ways to earn or borrow the required funds.

For beginning pilots who are making the airlines their goal, FAPA recommends the following progression in flight experience: Private Pilot rating, Commercial Pilot rating, Instrument rating, Certified Flight Instructor (CFI), Certified Flight Instructor Instrument rating (CFII), Multi-Engine rating (ME), Multi-Engine Instructor rating (CFIME), Flight Engineer written examination, Air Transport Pilot written examination, ATP rating, Flight Engineer rating (FE), and finally type rating in jet and heavy aircraft.

Building flight time and experience is the key to climbing the aviation employment ladder. The faster you earn your commercial and instrument ratings, the faster you can become compensated for your time. Up until that point, it's usually an out-of-pocket expense. The certified flight instructor rating is a valuable tool in the aviation industry. It is the foundation of training for every hopeful pilot. As you train new pilots, you can log much of that same time as pilot-in-command (PIC) time and be compensated for doing so. However, it is a job that requires skill, patience, and determination.

Moving up the ladder, the experienced CFI typically joins the ranks of heavier and more complex aircraft operators. Some pick up the additional pilot ratings along the way that are sought after by most major airlines.

BRANCH OF MILITARY
1988 Major Airline New Hire Pilots

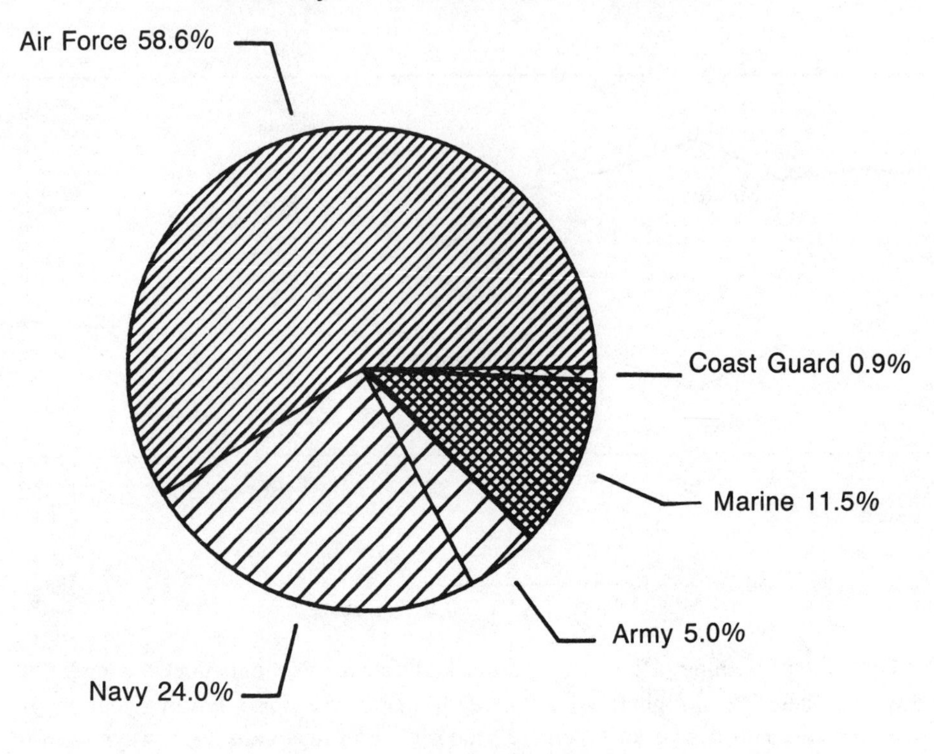

FAPA, Atlanta, GA 1-800-JET-JOBS

In 1987, over 80 percent of the new hire pilots selected had an ATP rating. Over 20 percent of the pilots hired had a FE rating, while the majority of pilots only had passed the FE written. About 90 percent of the pilots selected had a MEL rating, and approximately 50 percent of the pilots were type certified. It is also interesting to note that almost 40 percent of the pilots held a CFI rating.

Of these pilots hired in 1987, most had over 4,000 hours of total flight time. This number has been fairly stable over the past four years as shown in the graph.

Due to the lack of qualified pilots in recent years, the airlines have been forced to relax the number of required hours of jet experience. In 1981, the average pilot had about 2,500 hours of jet time, while at present that number is less than 2,000.

FACETS OF PILOTING

This section details the different kinds of piloting jobs and the features of each, including working conditions, opportunities for advancement, and future outlook. This information is courtesy of the FAA. Aviation Career Series pamphlet GA-300-122-84.

FLIGHT RATINGS & CERTIFICATES
1988 Regional New Hire Pilots

Percent

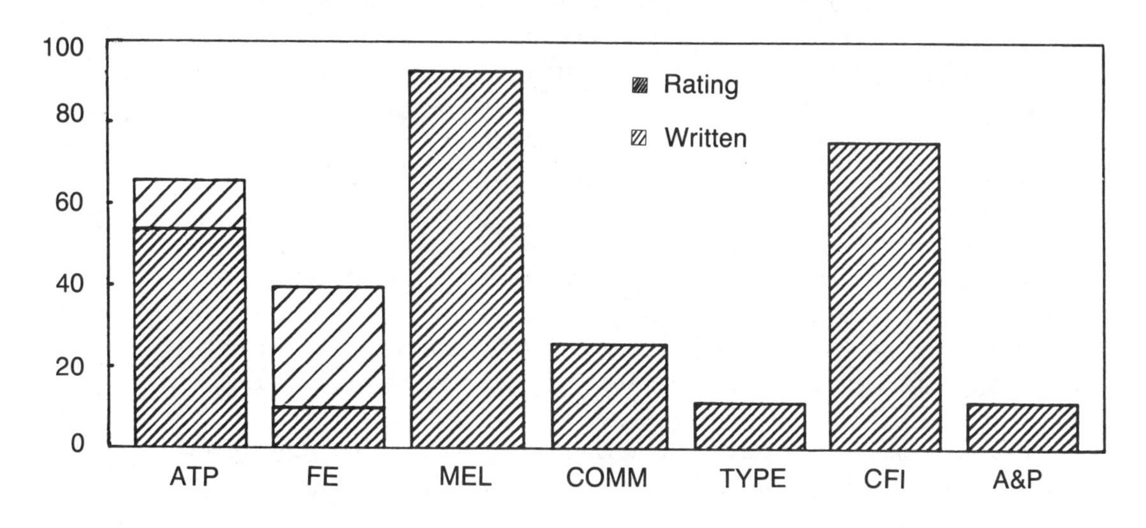

FAPA, Atlanta, GA 1-800-JET-JOBS

While the various kinds of piloting jobs involve a variety of special circumstances, there are also a number of conditions that are common to all pilots.

- All pilots flying for hire have progressed through a flight training program and have earned a commercial pilot's license and/or an airline transport rating. Most likely they also have one or more advanced ratings such as instrument, multi-engine, or aircraft type ratings, depending upon the requirements of their particular flying jobs.
- A pilot's "office" is the cockpit, which contains all controls, instruments, and electronic communication and navigation equipment necessary to operate the aircraft. Some noise and vibration are noticeable, particularly in propeller aircraft.
- They have a concern for safety including the safe condition or airworthiness of the plane; weather factors affecting the safety of the flight; flight regulations; air traffic control procedures, and air navigational aids designed to provide maximum safety in the air.
- Pilots also have a dual responsibility. They must not only satisfy their employer, who might be an air taxi or an airline operator, but they must also demonstrate to the Federal Aviation Administration (FAA) that their flying skills, knowledge, and state of health are at all times acceptable for the particular flying jobs they perform.

- They must undergo frequent physical examinations and meet certain medical standards that vary according to the license the pilot holds. A Class I Medical Certificate requires the highest standards for vision, hearing, equilibrium, and general physical condition. The pilot must have an exceptionally good health history with no evidence of organic and nervous diseases or mental disorders. A Class II Medical Certificate is less rigid but still requires a high degree of physical health and an excellent medical history. A Class III Medical Certificate has the least stringent physical requirements. All three classes of medical certificates allow the pilot to wear glasses provided the correction is within the prescribed limits of vision. Drug addiction and/or chronic alcoholism disqualify any applicant.

The greater the number of flying hours and the more complex the flying skills, the more varied are the opportunities for advancement as a pilot. There are many chances to transfer from one kind of pilot job to another as flying hours are accumulated and additional skills are mastered. Frequently, pilots double as flight instructors and air taxi pilots, or they might also operate an aircraft repair station with flight instruction and air taxi operations as sidelines.

The average major airline pilot earns approximately $70,000 per year, while regional airline/commuter pilots average $20,000 per year as do some charter and air taxi pilots.

AVERAGE FLIGHT EXPERIENCE IN HOURS

Major Airline New Hire Pilots 81–88

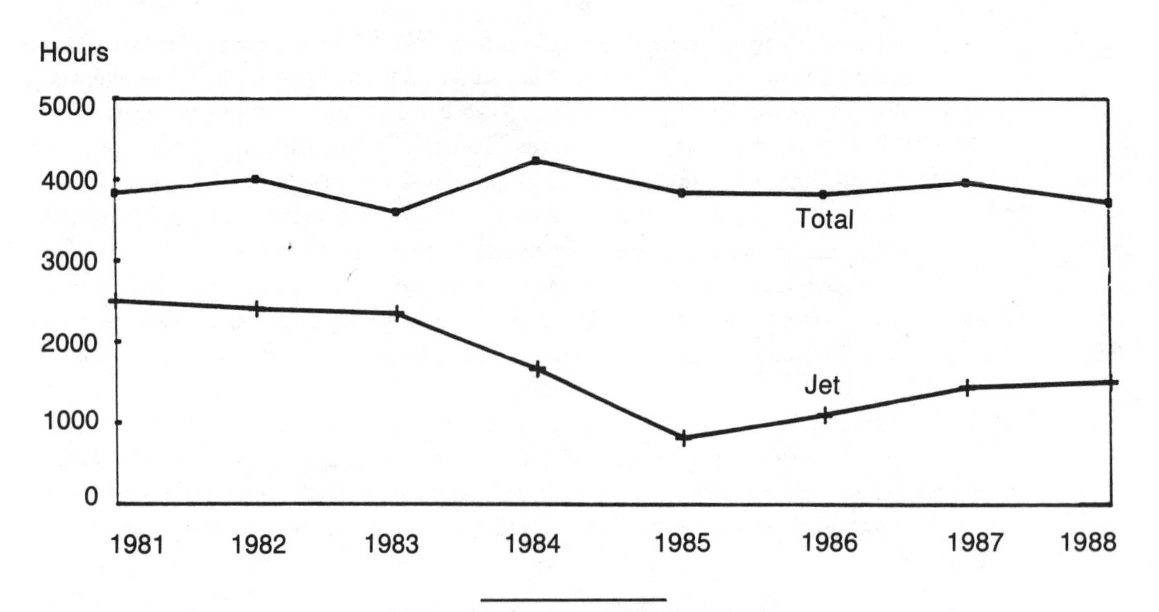

FAPA, Atlanta, GA 1-800-JET-JOBS

Helicopter pilots who fly to offshore oil rigs average salaries of $30,000 per year. These salaries vary significantly based upon the individual's age, education, work experience, and flying background.

FLIGHT INSTRUCTOR

Nature of the Work. Flight instructors teach students to fly. They demonstrate and explain the basic principles of flight, aerial navigation, weather factors, and flying regulations both on the ground and in the air. They demonstrate operations of aircraft and equipment in dual-controlled planes. They observe solo flights and determine each students' readiness to take examinations for licensed ratings. They also assist advanced students in acquiring advanced ratings such as commercial, instrument, multi-engine, and air transport ratings.

Working Conditions. Hours of work are irregular and seasonal depending on students' available time and the weather. Flight instructors might work as many as 80 hours a week during the summer and can expect to work every weekend with good flying weather throughout the year. The ground school classes might be scheduled during evening hours. Instruction duties rarely require being far from home base. When not teaching, flight instructors can supplement their income by working as an air taxi pilot or operating an aircraft repair station.

Where the Jobs Are. About 57,000 women and men in the U.S. hold flight instructor ratings, but only 12,000 are actively working. They usually fly from airports that have general aviation aircraft repair stations or an air taxi service where the operator provides flight instruction as an additional source of income. Flight instructors in areas with major airports having heavy air traffic usually operate out of the smaller airports in the community so beginning students can avoid heavy air traffic patterns.

Opportunities for Advancement. The job of flight instructor often is considered a stepping stone to higher paying flying positions. Thus, there is a large turnover in personnel and job openings. It should be noted that certified flight instructors (CFI) work long and irregular hours for low pay, $8.00 per hour, or less than $10,000 per year. Flight instructors who accumulate the necessary flight hours and experience often move on to jobs as corporate or airline pilots, but some remain in the teaching field. If they attain certain high standards, they can qualify for the FAA's Gold Seal, which identifies them as superior teachers and can lead to higher salaries. When the number of students is large enough, a flight instructor might organize a flying school, directing the activities of a number of instructors.

Outlook for the Future. General aviation is presently experiencing little or no growth in the number of people who want to learn to fly. The recent recession combined with elimination of the G.I. Bill Flight Training Benefits has had a detrimental effect on civilian pilot training. In the long run, today's general aviation fleet of approximately 220,000 aircraft is expected to increase to 315,000, creating a demand for more pilots and increasing the opportunities for flight instructors.

Corporate Pilot

Nature of the Work. Corporate pilots fly aircraft owned by business and industrial firms, transporting company executives on cross-country flights to branch plants and business conferences. They arrange for in-flight passenger meals and ground transportation at destinations and are responsible for supervising the servicing and maintenance of the aircraft and keeping aircraft records.

Working Conditions. The job is often demanding, but challenging, as the pilot is expected to fly in all kinds of weather into many unfamiliar airports. The aircraft could be a light twin-engine plane, a small executive jet, or even an airline type. The pilot is at the call of company executives, so he or she is subject to irregular hours. Often the pilot is away from home overnight. (Studies show that a significant percentage of round trips are over 1,000 miles.) If the company owns a fleet of planes, pilots might fly a regular schedule. Compared with the airline pilot, corporate pilot flying assignments are far from routine.

Opportunities for Advancement. A corporate pilot can acquire enough flight experience and skill on the job to qualify as an airline copilot. If the pilot prefers to remain in general aviation and the firm has a fleet of aircraft, she or he may eventually move up to the position of Chief Pilot, who directs all the aircraft operations of the firm.

Outlook for the Future. Studies of the growth of the business aircraft fleet indicate an accelerating interest in corporations owning aircraft in the years ahead. Over 500 of the top 1,000 companies in the U.S. have active flight departments. The advantages offered to business executives in time saving, privacy, and flexibility of schedules, plus improved aircraft that is specially designed for business use, are two important factors in the utilization of 50,000-plus company-owned planes in 1983. In 1983, business aircraft represented about 23 percent of all general aviation; however, they did approximately 75 percent of the general flying. General aviation activity amounted to 76 percent of the total aircraft operations at airports with FAA airport traffic control towers. To operate this expanding fleet requires about 1,500 new pilots each year, not including additional pilots to replace those who retire, transfer, or are removed for other reasons. Companies are expected to be in competition with the airlines in the hiring of qualified pilots, most of whom will be instrument rated.

Air Taxi (Charter) Pilot

Nature of the Work. An air taxi or charter pilot flies fare-paying passengers ''anywhere, any time,'' but usually for short trips over varying routes in single-engine or light twin-engine planes.

Working Conditions. These pilots fly passengers and cargo as service demands, but normally in daylight hours if the aircraft is a single-engine plane. Flights are mostly of short duration, and pilots can count on returning home at the end of the working day. If the pilot works for a company with a fleet of aircraft, she or he might fly on regular schedules over the same routes, much like a small airline. Pilots might be required to wear a uniform when on duty.

Where the Jobs Are. Air taxi operators are located at major airports and at other airports where sufficient passenger traffic can be generated. Interline agreements with airlines account for a substantial part of air taxi business; therefore operators are frequently located at airports having airline service.

Opportunities for Advancement. As is the case with the flight instructor, the air taxi pilot can build up enough flight experience in a relatively short time to qualify for the position of corporate pilot or air transport copilot. If the pilot elects to remain in the air taxi and charter business, he or she can generate enough business to offer "commuter airline service" or scheduled service over specified routes similar to the operation of a small airline.

Outlook for the Future. Air taxi and commuter operators claim the fastest rate of growth among all segments of general aviation. This growth reflects the increase in airline travel and the increased use of air taxis to "fly all the way" from any of the more than 400 airports in communities without airline service. Many airlines have agreements with air taxi companies to promote the use of air taxi service to airports not served by the airline and to issue through tickets. It also reflects a growing desire by the air traveler to bypass crowded metropolitan streets and use air taxis to reach destinations in outlying areas rather than rented cars. Since the Airline Deregulation Act was passed in 1978, there has been rapid growth in the air taxi and commuter industry. As the major airlines abandon unprofitable route segments, air taxi and commuter services move in to continue the necessary air service. Also, the U.S. Postal Service's practice of contracting with air taxi operators to deliver mail will further increase scheduled air taxi business. Given the present rate of expansion in this field, the need for air taxi and commuter pilots will continue to grow at a high rate.

COMMERCIAL AIRPLANE OR HELICOPTER PILOT

Nature of the Work. The commercial airplane or helicopter pilot performs a variety of flying jobs. If piloting a fixed-wing plane, the pilot might engage in such flying jobs as aerial photography, aerial advertising, sightseeing, geological survey, fish and game census, highway patrol, or checking federal airways and navigational aids. Helicopter pilots might fly on a regular schedule carrying workers and supplies to offshore oil rigs, fly accident victims to a hospital heliport, lift heavy loads to tops of buildings or to remote mountain sites, rescue people stranded by floods, carry smoke jumpers to fight forest fires, or deliver Santa Clause to shopping center parking lots.

Working Conditions. Flights are usually of short duration. The pilot usually works for an operator whose services are chartered. Helicopter pilots are often required to do precision flying hovering over a particular spot or landing in areas regular airplanes can't.

Where the Jobs Are. As the use of general aviation aircraft and helicopters is so varied and widespread in the U.S., pilots are employed just about everywhere there are airports or heliports.

Opportunities for Advancement. These pilots can aspire to advanced status as they build up hours of flying experience and skills. If they work for an operator who owns a fleet of aircraft or helicopters, they could advance to the job of Chief Pilot or build up enough business to employ other pilots and direct their operations.

Outlook for the Future. At best, the outlook is mixed for the short term considering the recent recession, rising fuel prices, and decrease in aircraft production. Studies do indicate that for the long term (1988 through the 1990s), the need for pilots will grow as more pilots retire and the demand for the various aviation services grows.

PATROL PILOT

Nature of the Work. The patrol pilot flies cross-country at low altitudes along pipelines or power lines, checking for signs of damage, vandalism, and other conditions that might require repairs. Patrol pilots radio to headquarters the location and nature of repair jobs.

Working Conditions. Patrol pilots fly light aircraft over all kinds of terrain, frequently at low altitude. Many jobs are working for an operator who contracts with an oil pipeline or electric power company to furnish aerial patrol service.

Where the Jobs Are. Patrol pilots fly wherever electrical power transmission lines or oil and gas pipelines exist. Many power transmission lines run through mountainous regions where water sources and dams produce electrical power. Oil and gas pipelines spread out in underground networks from oil and gas fields, many of which are located in midwestern and southern states. Some pilots are employed by U.S. Immigration Service to patrol the international borders.

FERRY PILOT

Nature of Work. The ferry pilot flies new aircraft from the manufacturing plant to dealers' showrooms and to private customers' home airports.

Working Conditions. After delivering new aircraft to customers and dealers, the pilot returns to her or his home base on a commercial airliner or by another form of transportation. The pilot may be away from home overnight, depending on the distance required by the ferry flight. Ferry flights may required flying to foreign countries.

Where the Jobs Are. Ferry pilots often operate out of cities that have light-aircraft manufacturing plants, most of which are concentrated in Kansas, Oklahoma, Florida, and Pennsylvania.

Outlook for the Future. The production of general aviation aircraft has decreased from 17,000 in 1978 to approximately 3,000 in 1983. These recent reductions in production have significantly lessened the need for ferry pilots to deliver new aircraft. The active used aircraft market has increased demand somewhat, but it is still quite low.

AGRICULTURAL PILOT

Nature of the Work. The agricultural pilot sometimes called an *aerial applicator* or "crop duster," flies specially designed aircraft (including helicopters) to apply herbicides, insecticides, seeds, and fertilizers on crops, orchards, forests, fields, and swamps. Some jobs require aerial surveys of cattle and crops or fighting forest fires by dumping fire-retardant materials.

Working Conditions. These pilots fly at low altitudes with heavy loads, in a regular pattern over the ground, avoiding trees, power lines, fences, and other obstacles. Most flying is done during the early hours of the morning and again in early evening when the air is still. Takeoffs are often made from country roads and open fields close to the area to be treated. Work is seasonal, ranging from six to nine months in southern areas to two months in northern sections. The operator usually furnishes the aircraft, trained ground crews, and specialists who decide how the land is to be treated. The pilot works very close to poisonous liquids and chemicals and must wear protective clothing and masks.

Where the Jobs Are. The agricultural pilots are in demand mostly in California and in the southern tier of states where the crop growing season is at its longest. Many pilots follow the crops north as the season progresses, while others find work in northeastern and western states with extensive forest areas.

Outlook for the Future. The number of agricultural operators in the U.S. has grown to 3,300, employing more than 25,00 people and operating some 9,000 aircraft that make applications to more than 180 million acres of farmland each year. Experienced agricultural pilots continue to be in demand.

TEST PILOT

Nature of the Work. Experimental or engineering test pilots fly newly designed and experimental aircraft to determine if the plane operates according to design standards, and they can make suggestions for improvements. Production test pilots fly new planes as they come off assembly lines to make sure they are airworthy and ready to turn over to customers. Airline test pilots flight test airliners after major overhauls before the planes are put back into service. They also flight test new aircraft to be sure they are up to airline standards before the airline accepts them from the manufacturer. Test pilots for the FAA fly FAA planes with experimental equipment aboard to test performance of the equipment, or they fly FAA planes to test new kinds of ground-based navigational aids such as radar or runway lighting.

Working Conditions. The test pilot's job involves the most flying hazards. The experimental test pilot must expect the unexpected as the plane is tested to the limits of its design strength and performance capabilities. The production test pilot tests a plane on the basis of expected performance and known standards, as does the airline test pilot. All of these pilots sometimes encounter emergency situations that they are expected to handle with the skill and knowledge their job requires. They prepare written and oral reports on their flight experiences and can fly either during the day or at night, depending upon the requirements of the test flight. Airline test pilots often work at night or on weekends, as most aircraft are serviced at that time.

Where the Jobs Are. Experimental and production test pilots are employed at all aircraft manufacturing plants, which are located mainly in California, Washington, Kansas, Texas, Georgia, Oklahoma, Maryland, Missouri, Florida, New York, Pennsylvania, and Connecticut. Airline test pilots work wherever the airlines have overhaul bases, the largest ones of which are found in San Francisco, Miami, New York, Tulsa, and Kansas City.

Opportunities for Advancement. Engineering test pilots may advance to the position of Chief Test Pilot, as can production test pilots. Airline test pilots eventually may advance to the airline's engineering or maintenance administrative staff. Test pilot jobs are also available with the FAA.

Outlook for the Future. The demand for engineering and production test pilots will fluctuate with the development and production of aircraft. Over the next decade the production of general aviation aircraft is expected to increase, while that of commercial air transports will level off due to the introduction of the jumbo jets.

AIRLINE PILOT (CAPTAIN)

Nature of the Work. The airline pilot plans each flight with the airline's flight dispatcher and meteorologist, checking weight, fuel supply, alternate destination, weather, and route. The pilot also briefs the crew, checks out takeoff procedures, ascertains that the plane is operating normally before takeoff, gets takeoff clearance from the air traffic control tower, flies the plane over the designated route, lands the plane, and at the final destination files a trip report. During the time the airline pilot is aboard the aircraft, he or she supervises the work of the crew, gives instructions, and makes all decisions. The captain is in command of the plane and is responsible for the safety of the aircraft, its passengers, crew, and cargo. The aircraft flown can range from a twin-engine DC-3 on a 100-mile hop to a four-engine Boeing 747 crossing the ocean.

Working Conditions. By law, an airline pilot may not fly more than 85 hours a month or 1,000 hours a year. However, the average pilot works more than 100 hours a month counting ground duties such as filing flight plans, working on reports, briefing crews, and attending training classes. The airline pilot spends most of the working day in the cockpit with additional time in the airline dispatcher's office and in training classrooms. Work schedules average sixteen days a month and usually provide for consecutive days off. Schedules for pilots employed by transcontinental and international airlines require pilots to spend some nights away from home. In these cases, hotel, transportation, and meal expenses are paid by the airline. A flight requires considerable pilot concentration during takeoff and landing maneuvers. Automatic piloting devices free the pilot for other cockpit duties and lessen the strain of the job during cruising flight. The airline pilot is required to wear a uniform while on duty. Night flights are often required, especially for air cargo operations.

Where the Jobs Are. Scheduled airline flight crews are based at major terminals on their respective airline routes. These bases are found mainly in New York, Chicago, Los Angeles, San Francisco, Seattle, Detroit, Newark, Atlanta, Miami, District of Columbia, Denver, Dallas, and Cleveland. Flight crew job opportunities are also available with all cargo airlines and with nonscheduled and supplemental airlines that provide charter service.

Opportunities for Advancement. Promotion is regulated by seniority. When hired as a second officer, or copilot, the person is assigned the bottom position within the airline. Copilots and pilots advance to larger aircraft and retire, resign, or are removed from the list for other reasons, and the newly hired pilot moves upward. All through the career with the airline, the earnings, route assignments, and vacation time preferences are governed by seniority. Second officers of flight engineers can advance to co-pilot position within a year, but it usually takes from seven to twelve years to become a pilot or captain, depending on the size of the airline and rank on the seniority list. By law, pilots must retire when reaching age 60, but flight engineers can fly up to the age of 65. All through the pilot's career, he or she must ''lay the job on the line'' every six months at the time of a rigid physical exam. If unable to pass the physical, the pilot must stop flying.

Outlook for the Future. The outlook for career opportunities for pilots and flight engineers with the airlines is directly related to airline growth. Airline growth is usually measured by an increase in traffic, i.e., an increase in passenger-miles and an increase in ton-miles of freight. This growth, of course, is directly related to the health of the national economy. The recent recession (1979–1982) combined with airline deregulation has had a detrimental effect on airline hiring, but the future looks promising.

Airline Copilot (First Officer)

Nature of the Work. The airline copilot or first officer assists the captain by monitoring the flight instruments, handling radio communications, watching for air traffic, and taking over the flight controls when directed by the captain.

Working Conditions. Approximately the same as for the airline captain.

Where the Jobs Are. Approximately the same as for the airline captain.

Opportunities for Advancement. Can move up to airline captain with seniority.

Outlook for the Future. Approximately the same as for the airline captain.

Flight Engineer (Second Officer)

(The latter title applies when the employee is required to have minimum training as a copilot.)

Nature of the Work. The Flight Engineer makes a walk-around inspection of the aircraft, checking approximately 200 items. She or he oversees fueling operations, reviews mechanics' reports, and assists the captain with preflight cockpit check. He or she also monitors engines, keeps track of fuel consumption and the heating, pressurization, hydraulic, electrical, and air conditioning systems. Flight engineers or second officers troubleshoot and, if possible, repair faulty equipment in flight, check and maintain aircraft log books, report mechanical difficulties to mechanic crew chief, and make a final post-flight inspection of the aircraft.

Working Conditions. Work schedules are approximately the same as for the other pilot categories.

Where the Jobs Are. Approximately the same as for the airline captain.

Outlook for the Future. Approximately the same as for the airline captain.

TRAINING OPPORTUNITIES FOR PILOTS AND FLIGHT ENGINEERS

There are several approaches to acquiring pilot training. The first is through flight instruction at FAA certificated flying schools. The student must be at least 16 years of age and be able to pass a third class medical examination. Courses consist of 40 hours of ground school instruction where students learn the principles of flight, aerial navigation, weather factors, and flight regulations. Forty hours of flying lessons are conducted in dual-controlled aircraft (20 hours dual instruction and 20 hours solo flight). The instructor judges when the student is ready to take the written and flight examinations, which are given by FAA inspectors. Upon successful completion of both exams, she or he earns the private pilot's license that entitles the pilot to fly passengers, but not for hire.

The private pilot can then undertake advanced instruction to learn to fly on instruments to obtain an instrument rating. The next step is to earn a commercial pilot's license with additional hours of flight training and experience. These achievements open financial doors, because now the pilot can fly for hire. Further study and experience could eventually earn him or her the Air Transport Rating for qualification as an airline pilot.

A second method of acquiring flight training is through pilot training in the armed forces. This entails no expense to the student other than a five-year service obligation. With some additional study, the military pilot can qualify for numerous civilian pilot jobs upon leaving the service. The military services have been a major source of pilots for the airlines.

Thirdly, a growing number of colleges and universities offer flight training with credit toward a degree. The graduate leaves school with a private, commercial and/or certified flight instructor license, and in a few cases, an Air Transport Rating plus a degree.

Helicopter pilots can receive training in the armed forces or at special private FAA certificated helicopter flight schools. Agricultural pilots can receive specialized advanced training at agricultural pilot schools.

Some airlines offer training courses for corporate pilots transitioning to new jet aircraft. The airline's experience in jet flight training makes them particularly well qualified to provide this service to business firms.

See Chapter 2 for more information on the FAA-approved two-year, four-year, and trade schools that offer every phase of flight training.

Career Profile

Professional Pilot

Name: Rick C. Bogatko II

Career Field: Professional Pilot

Position Held: First Officer (Copilot), Boeing 727
Second Officer (Flight Engineer), Boeing 747
Northwest Airlines

Responsibilities

General: All Flight Officers are responsible for the safety of passengers, crew members, cargo and airplane. They must maintain a valid airman's license and medical as required by the Federal Aviation Regulations (FAR's). They must be thoroughly familiar with and comply with all FAR's pertinent to their licenses and air carrier operations. The *only* deviations from FAR's and company operation specification are when the safety of the flight necessitates a different course of action. Flight Officers must maintain a professional image in conduct and personal appearance at all times while in uniform.

First Officer: As a First Officer, responsibilities initially include a review with the captain of the proposed route of flight, the reported weather, the forecast weather for the departure, all enroute, destination, and alternate airports, and to verify with flight dispatch that the flight can be conducted safely. The First Officer also copies ATC enroute clearances, taxi clearances, and handles communications and navigation as well as flying duties.

Second Officer: As Second Officer, initial responsibilities include assuring that the aircraft is mechanically safe for the proposed flight. This includes checking maintenance log books, checking all systems on board the aircraft (both inside and outside), and assuring that all necessary servicing, fueling, oil, oxygen, hydraulics, potable water, spare light bulbs, forms, etc. has been properly accomplished. The Second Officer is also responsible for checking weather reports and forecasts and route requirements, while the Captain and First Officer are responsible for performing specific aircraft systems checks prior to flight. Only when all crew members have all available information concerning the proposed flight can they finally operates as a single unit—a crew.

Education

All major air carriers prefer four years of college. My B.S. degree is in Conservation with a minor in Wildlife Management. Aviation is not a secure enough industry to be totally dependent on, so it is a very good idea for a pilot to have a second career and education to fall back on, should you lose your medical, be furloughed, or fail a check ride.

I began flying in junior high school when I earned the aviation merit badge in Boy Scouts. I got my private pilot's license while still in high school. While in college, I concentrated on my undergraduate studies and maintained my flight currency. After graduation I moved to Phoenix, Arizona—an environment I felt would offer me the best opportunity to earn as many flight ratings and as much flight time as quickly as possible. I took a job with a fixed-base operator (FBO—a flight school and charter flight business) as a ramper (I washed, waxed, and fueled aircraft) and worked in the maintenance shop. I simultaneously worked on my commercial, instrument, certified flight instructor, instrument instructor, multi-engine land, multi instructor and airline transport pilot ratings.

After becoming a CFI, the FBO put me on the instructors staff. I was finally earning a living from piloting aircraft. As an instructor, I was the pilot in command and logged all the same flight hours as my students. I eventually had logged enough time to start flying FAR part 135 charter flight in addition to instructing, moving from large single-engine aircraft to twins. I earned all my flight ratings within 18 months.

I worked as an FAA Accident Prevention Counselor for 2 years and I earned the Gold Seal Flight Instructor Certificate. In 1983, I was honored by my students, peers, ATC specialists, and FAA examiners by being awarded Flight Instructor of the Year for the state of Arizona.

At the same time I was instructing, I was investigating my chances of being given a commission to fly in the military. Unfortunately, due to the political environment at the time, the government was cutting back on military spending, and flying commissions were very difficult to get. In the winter of 1983, I finally located three commissions to fly tankers in the Air National Guard. While testing for these military commissions, I successfully interviewed and was offered a civilian position with a regional airline as a Beechcraft C-99 copilot. Hence, I made the decision to pursue the civilian route to the majors. After three months, I upgraded to Swearingen Metroliner Copilot, and after one full year, upgraded to Metro Captain. I spent one year as a Captain, and two years with the regional before being offered a position with a major.

In my ''new-hire class'' at the major airline, I was the youngest at 28 (with the oldest being 36), and I was the only strictly civilian pilot out of a group of 14. I also had more than twice the flight time, at 5200 hours, than the next pilot in my class.

Benefits and Drawbacks of the Job

When scheduled to fly anywhere from one to fourteen days with two other crew members and you know you will be working in a confined space with them for as long as 15 hours at a stretch, you must be able to communicate and get along. You can't allow your ego or pride to interfere with getting the job at hand done safely and efficiently, so it helps to have an amiable personality.

Being flexible is another necessity for the job. The airline changes flight schedules on a seasonal basis. This changes the staffing requirements among the various pilot bases and on the various types of planes flown. Although I live in Seattle and usually fly the 747 out of that city, it has been necessary for me to fly out of Minneapolis and New York for months at a time, and I also must change from the 747 to the 727 and back. It's all part of the job.

Another drawback is being the junior person in a given position on a given piece of equipment in a given base when you first start out. For example, as a newly hired pilot, you will most likely be trained on the most ''junior'' airplane (i.e.; 727 vs. the 747) in the most junior position (Second Officer vs. Captain) at the most junior base (Detroit vs. Boston). Once on the flight line and in a base, you will gain seniority relative to the position, the equipment, and the base. (Should you elect to remain in the junior base and on the junior equipment, you can become the ''senior junior'' person in that area.)

On the other hand, being junior usually means that your base is assigned and there is no schedule left available when it is your turn to bid. So, you sit on reserve 24 hours a day on certain scheduled days of the month. You must be able to report for flight duty within two hours of being notified that you've been assigned a trip. You usually have to fly over all the holidays, and you might not be able to go home for birthdays, graduations, or weddings. You could end up having to take vacation when your kids have to be in school, and when you *can* get a trip, it's entirely possible you would have to leave Friday and get home Tuesday with overnights in Hoboken, Bemidji, and Sioux Falls.

Seniority entitles you to *bid*, or choose, your flying position and trip schedule. As the most senior person, you can now elect to fly only those schedules that for example leave on Mondays at noon and return to base on Thursdays at 2 p.m. with layovers in Miami on Monday night, Phoenix on Tuesday night and L.A. on Wednesday night. Being senior allows you to bid to be home over the holidays and special occasions or for the most desirable time of year for vacations. Eventually you can bid for the positions of First Officer and Captain, for larger, more desirable aircraft, and to get into your preferred base city. Therefore, through attrition, expansion, and patience, everyone can look forward to eventual seniority and flying those 747 trips to Hong Kong, Bangkok, Frankfurt, and other fascinating places as a Captain.

Another benefit to a flying job is the longer period of time you can spend at home. Flight crew members might be away for 3 to 14 days at a time, but they are home for 3 to 14 days at a time. Although the total number of hours spent away from home per month is often more than for someone on the normal 8-hour, 5-day work week schedule, the hours spent at home are higher quality time because you don't have the ''interruption'' of a daily routine.

Flying isn't just a career choice—it's a way of life. It's a way of life that I have wanted for as long as I can remember, and I wouldn't want to trade it for any other.

FLIGHT ATTENDANT

Flight attendants provide two basic services during flight operations on most passenger-carrying aircraft. Stated simply, the flight attendant assures the safety and comfort of passengers during flight. Flight attendants are usually briefed by the aircraft captain on such things as expected weather conditions, special passenger problems, and other matters relative to the flight. It is the flight attendant's job to ensure that adequate supplies of food, beverages, reading materials, blankets, and pillows are aboard the aircraft. Inspection must also be made of first aid kits and other emergency equipment to verify that they are in satisfactory condition in case they are needed. When boarding the aircraft, flight attendants check passenger tickets, answer questions on passenger seating, and assist in storing any carry-on luggage.

Before the flight departs, attendants give instruction on the use of emergency equipment and procedures. Attendants are trained to administer first aid to passengers when needed. While airborne, the flight attendants serve refreshments, snacks, and on extended flights, a prepared supper is provided. After landing, attendants provide assistance when deplaning the aircraft.

Because airlines operate around the clock, all year round, flight attendants often work staggered shifts, both day and night as well as on holidays and weekends. Many attendants spend a considerable amount of time away from their home bases. During this time, the airline provides hotel and meal allowances.

According to the U.S. Department of Labor, flight attendants held 64,000 jobs in 1984. Commercial airlines employed the majority of these attendants, while a small portion were employed by large companies who operate their own aircraft for business travel. According to the Future Aviation Professionals of America (FAPA), airlines hired more flight attendants in 1986 than in previous years. This hiring trend is expected to continue because of the increasing demand for airline travel. FAA safety rules require one flight attendant for every fifty seats on the aircraft. Therefore, with the increasing number of large jets being built, there is a growing demand for additional flight attendants.

It is the airline's goal to create a well-rounded, customer-oriented attendant who represents the organization in a professional manner. An applicant preparing for a career as a flight attendant can greatly improve his/her chances of acceptance if he masters the use of the English language and speaks clearly using proper grammar and poise. Although not mandatory on most assignments, many airlines like to see applicants who are bilingual.

Applicants must be high school graduates, and most airlines like to see a few years of college as well. Many airlines have weight, height, and age requirements for flight attendants. For specific requirements, applicants should consult the airlines of interest. Excellent health, good vision, nice teeth, and a neat, well-groomed appearance are essential qualities for flight attendants.

All airlines regard public contact work such as working for hospitals or charitable organizations as invaluable in preparation for a career as a flight attendant. To the prospective employer, volunteer work demonstrates responsibility, dependability, and reliability in the absence of work rules and compensation. It also shows a sincere desire on the applicant's part to work with other people. The successful applicant must be able to deal with people in a pleasant and efficient manner, no matter how tiring or stressful the situation may be.

At present, most airlines require newly hired flight attendants to attend a four- to six-week training seminar. These training seminars typically provide the applicant with a basic understanding of his/her job responsibilities. Topics include proper dress and grooming, an introduction to food and beverage service, public relations procedures, emergency first aid, and an overall familiarization with aircraft emergency procedures including forced landings, evacuation procedures, decompression, and ditching.

FLIGHT ATTENDANT DEMAND

FAPA, Atlanta, GA
1-800-JET-JOBS

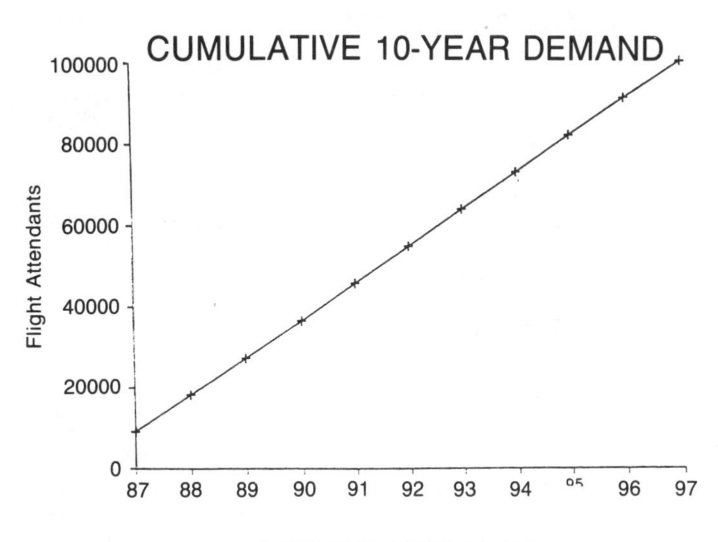

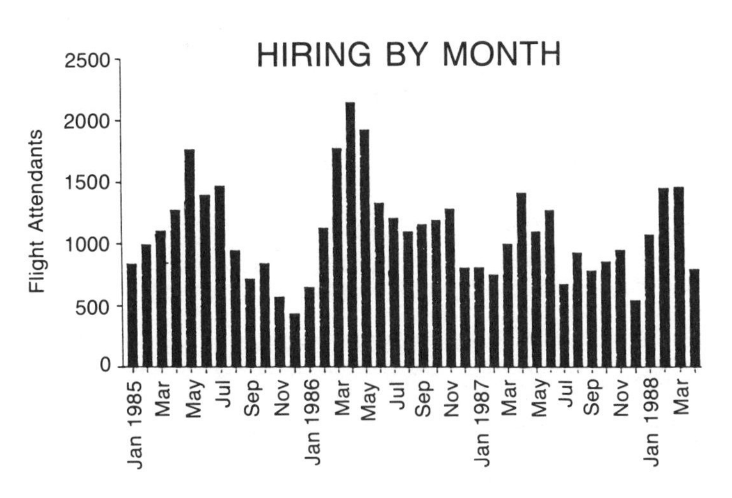

Career Profile

Flight Attendant

Name: Karen Lee
Career Field: Flight Attendant
Position Held: Flight Attendant, Alaska Airlines
Education: Bachelors Degree in Physical Education
University of Minnesota at Duluth
Professional Affiliations: AFA (Association of Flight Attendants)

Experience

Since high school, all of the jobs I have had were customer-service oriented and involved public contact. After college, I was hired by National School Studios and transferred to Portland, Oregon to work as a school photographer and sales representative. After three years on the job, I moved to Seattle, Washington and began working at Nordstrom, a Seattle-based retail apparel store. Nordstrom is known for its excellent customer service. I worked in the customer service department for two years and in the data processing department for one year. Because I enjoyed working for a customer-service-oriented company, when it came time for me to apply to the airline, I was looking for an airline with similar goals and standards as Nordstrom. Alaska Airlines was my choice because they are also well known for their excellent customer service. My work experience at Nordstrom probably prepared me the most for work as a flight attendant because of the exposure to many different kinds of people and experiences.

Position Responsibilities

A flight attendant is responsible for many different functions. The following is a list of the most important ones:

- Passenger safety is the number-one priority. The flight attendant must assume complete charge of the cabin and passengers in an emergency situation.

- The flight attendant must see that all passengers observe safety regulations such as stowing carry-on items, buckling seatbelts, etc.
- Flight attendants must attend a preflight briefing with the rest of the crew.
- They are required to observe all company policies, procedures, and uniform rules and regulations.
- They must check aircraft cabin emergency equipment and catering prior to departure.
- Flight attendants ensure all announcements are made as required by Federal Aviation Regulations and company policies.
- Flight attendants ensure that all passengers deplane at the proper exit upon destination arrival.

Benefits and Drawbacks of the Job

The best part of this job is the benefits. Airlines typically have excellent medical, retirement, vacation, and comp-time benefits.

In my opinion, the time off is the greatest benefit. A flight attendant has anywhere from 10 to 21 days off per month. This allows a person to pursue other interests, travel, or just relax at home.

The travel benefits are also excellent. In addition to free or very low air fares, airline employees receive discounts on hotels, rental cars, ski areas, and many other attractions (Disneyland, Sea World, etc.).

The flexible work schedule is also very attractive to a person who does not like the 8-to-5 routine. Flight attendants can trade trips with other flight attendants and by doing so can rearrange their entire schedule. Flight attendants can fly to all kinds of exotic places from warm and sunny resorts to the cold, icy tundra. A person needs a sense of adventure and must be prepared to be away from home for days at a time.

Layovers are almost like mini vacations. Layovers can be ten hours to several days in length depending on whether the trip is international or domestic.

Another positive aspect of being a flight attendant is that you meet all kinds of interesting people from all walks of life. This helps to keep the job exciting. Flying becomes a way of living—a lifestyle that is difficult to give up once you get accustomed to it.

However, some might consider the drawbacks of the job to be many. You really need to be prepared for the long hours and days away from home base. You have to be willing to relocate in some cases. Holidays might be spent in strange cities or far away places. The irregular schedule can cause extreme fatigue.

In addition to this, you have to be prepared to deal with difficult passengers, emergency medical situations, air sickness, turbulence, and other taxing situations. A flight doesn't always go as smoothly as planned, and you need teamwork to make the flight a success.

Flight attendants get colds and other sicknesses frequently. Anytime you work in close public contact with so many people, it is easy to pick up the latest germ strain on the market. Also, the changes in climate can fool your immune system.

For someone starting in the industry, the base pay is typically low. However, it is possible to earn *per diem* pay, incentive pay, international pay, and other extra earnings while performing the job. The more hours you fly, the more money you make.

A new hire always takes the risk of being furloughed. It is wise to get on with an airline that is financially secure and has a good reputation. In this industry, you must be prepared for anything. The name of the game in the airline industry is *seniority*. Those who have been in the industry longer are getting paid better and get the best choices of trips.

Another drawback of the job is being on reserve. A reserve flight attendant is on call 24 hours a day with the exception of 10 to 12 guaranteed days off per month. Each airline has its own reserve policy. Generally, the most junior flight attendants are the reserves. A flight attendant needs to be packed and ready to go at all times. During busy times, reserve flight attendants may only have a few hours previous notice to be at work. Prompt reporting is essential. This type of job is not for the individual who is always a few minutes late. It demands a certain amount of precision and dedication.

Recommendations to Job Hunters

Today, most airlines require a high school education followed by at least two years of college or two years of experience with the public. Other requirements include a pleasing personality and a neat, well-groomed appearance. Weight and height requirements vary from carrier to carrier. Some airlines are more rigid than others. Contact lenses and eye glasses are acceptable. Generally, most companies like to hire people of age twenty-one or older. I feel most airlines are in a trend of hiring mature, well-rounded individuals, many of whom have college degrees. Some positions require proficiency in a foreign language.

Once you do get your airline job, be prepared for five to six weeks of intense training. In this school, you learn about the different types of aircraft flown by the airlines, first aid, CPR, and emergency evacuation procedures. Other activities include viewing crash movies, practicing fire fighting, and learning hijack training and passenger psychology. Also included are procedures for briefing an unaccompanied minor or a blind passenger, the company's policies and procedures, Federal Aviation Regulations, and finally, food and beverage service training. To pass the training class, you need a good grasp of all the material presented. A flight attendant must attend and pass recurrent training every twelve months.

To sum it up, to prepare for a flight attendant position, you should first get a good education. Secondly, try to get as much public work experience as possible. Once your initial application for employment has been submitted, you may be called for an interview. The interview process involves several interviews and could include a group interview. My suggestion for the interview is to wear a suit if at all possible. A suit looks neat and professional and you are more likely to be remembered. Smile and be yourself; be confident. If you make it through the interviewing process, the next step is a physical examination. Then, if all is well, you're in.

Generally, a good flight attendant is independent, caring, versatile, flexible, dynamic, adaptable, has a good sense of humor, and is willing to work hard. According to various sources, the outlook for flight attendants is very favorable. If it's a career you're interested in, now is a great time to come on board.

FLIGHT OPERATIONS SUPPORT

There are many other airline professionals that work behind the scenes to make a passenger's transition from departure to destination as smooth as possible. In most cases, the quality of their work is as important as the flight itself. Much planning and preparation take place before the aircraft is cleared to taxi for takeoff. The following is a brief outline of some of these behind-the-scene employees.

METEOROLOGISTS

The meteorologist prepares reports on weather for flight personnel, airline operations, and traffic departments. These reports are used to determine the present weather conditions for flight planning purposes at the departure point, enroute and at the destination. No aspect can be as important as the safety of the aircrew and passengers. The meteorologist helps the aircraft steer clear of such hazards as thunderstorms, aircraft icing, and other related severe weather.

The successful applicant for a meteorological position with an airline should have at least a four-year degree (B.S.) in meteorology. Military weather service or U.S. Weather Bureau experience is a plus.

FOOD SERVICE COORDINATORS

Inflight meals have become a very important aspect in airline travel. Services range from inflight snacks such as peanuts to complete, pre-prepared dinners. Beverages are also served during flight.

The food service staff determines what type foods should be served on a particular flight. After the budget is set, the coordinators determine the quantity of food they need and then determine the quality of meals they can afford. Usually, airlines hire external caterers who prepare the food and deliver it in easily dispensable units. Much planning, budgeting, and coordination goes into serving the final product. The successful applicant for a food service coordinator's position should have an appropriate background in restaurant management.

CARGO HANDLERS

Cargo handlers load and service the aircraft during stops. It is the cargo handler's responsibility to load baggage, mail, commercial freight, food service supplies, or any other cargo necessary. Most such employees must operate machinery such as forklifts, hydraulic lifts, conveyor systems, and baggage transporting vehicles.

The successful applicant in this field has a high school education or equivalent and must be able to lift heavy items. Previous experience in this type work is desirable, but not necessary.

AVIATION MAINTENANCE

Aviation maintenance mechanics (including airframe and powerplant technicians, avionics technicians, and instrument repairmen) have the important responsibility of keeping airplanes in a safe condition to fly. In this effort, they service, repair, and overhaul various aircraft components and systems including airframes, engines, electrical and hydraulic systems, propellers, avionics equipment, and aircraft instruments. The nature of the work has changed greatly in recent years and will continue to change rapidly because of advances in computer technology, solid-state electronics, and fiber composite structural material. This information is courtesy of the FAA Aviation Career Series pamphlet GA-300-123-89.

material. This information is courtesy of the FAA Aviation Career Series pamphlet GA-300-123-89.

In General

Aircraft mechanics may be licensed or unlicensed. The licensed mechanic could hold a Mechanic's Certificate with an Airframe rating, Powerplant rating or both (expressed as "A&P"); or a Repairman's Certificate from the FAA. FAA certificates are issued upon successful completion of oral, written, and practical examinations. The Airframe, Powerplant, or Airframe and Powerplant Certificates allow a mechanic to work only on those specific parts of the aircraft; i.e., engines, airframe, and systems for which he or she is rated. The mechanic with the FAA's Repairman Certificate can work on those parts of the aircraft that the certificate specifically allows, such as radio, instruments, propellers, etc.

Aircraft mechanics employed by the airlines perform either line maintenance work (routine maintenance, servicing, or emergency repairs at airline terminals) or major repairs and periodic inspections at an airline's overhaul base.

Working Conditions. Depending upon the type of work they do, aircraft mechanics work in hangars, on the flight line, or in repair shops. They use hand and power tools along with test equipment. Noise levels are high and flight line mechanics often work outdoors in inclement weather conditions when making emergency repairs. Sometimes the work requires the use of ladders or scaffolds, and the physical demands can be heavy. Frequent lifts or pulls of up to 50 pounds are normal, and the physical requirements include stooping, kneeling, crouching, crawling, reaching, and handling.

Aircraft mechanics often work under pressure to maintain airline flight schedules, or in the case of general aviation, to minimize inconvenience to customers beyond a reasonable period of time. While doing so, the aircraft mechanic cannot sacrifice high standards of workmanship to speed up the job.

Where the Jobs Are. The scheduled airlines employ approximately 50,000 mechanics at various terminals and overhaul bases located throughout the U.S. and overseas. The major overhaul facilities are located in New York, Los Angeles, San Francisco, Miami, Denver, Atlanta, Kansas City, Tulsa, and Minneapolis.

In addition, approximately 85,000 A&P-licensed mechanics are employed in general aviation for air taxi and fixed base operators, aerial applicators, flight training schools, supplemental airlines, corporations owning fleets of aircraft, and aircraft manufacturers. Also, mechanics and technicians are employed at some 4,000 FAA-certified repair stations in the U.S.

Another large employer is the U.S. government, which employs approximately 100,000 civilian aircraft (certificated/uncertificated) mechanics and avionics technicians to work on military aircraft at Army, Navy, Marine Corps, and Air Force installations in the U.S. and overseas. In addition, FAA employs maintenance personnel who work at various locations in the U.S. and overseas. A majority of these persons work at the FAA's main overhaul base located in Oklahoma City, Oklahoma.

Wages and Benefits. Aircraft mechanics generally work 40 hours a week on 8-hour shifts around the clock, and overtime work is common. The basic airline mechanic's starting wage is approximately $24,000 per year ($12.00 per hour), but there are increases in salary for longevity, licenses held, line work, or shift work. A lead airline mechanic with an A&P certificate and 10 years experience can expect to make in excess of $40,000 per year ($20.00 per hour).

In general aviation, mechanic's salaries are determined largely by the size of the aircraft serviced. One national survey of general aviation mechanics holding an A&P license showed an average starting salary of $13,000 per year ($6.50 per hour) but increasing to $19,000 per year ($9.50 per hour) after five years on the job. Mechanics without an A&P license make considerably less and usually have more difficulty finding work. It is anticipated that wages for general aviation mechanics will increase over the next few years but will remain lower than the salaries paid by the large airlines.

Paid holidays, vacations, insurance plans, retirement programs and sick leave are some of the benefits offered by both airline and general aviation employers. Airliners also give their employees free or reduced price transportation to destinations within their route structure and exchange travel privileges with other airlines. General aviation offers more local points of employment.

The International Association of Machinists and Aerospace Workers and the Transport Workers Union of America are the principal unions representing aircraft mechanics, but some mechanics are also represented by the International Brotherhood of Teamsters.

Opportunities For Advancement. An apprentice mechanic who has gained the required experience with engines and airframes or an applicant who is a graduate of an approved aircraft mechanics course can acquire the A&P Mechanic Certificate. Mechanics with 30 months combined experience or 18 months airframe or powerplant experience may take the Airframe, Powerplant, or the Airframe & Powerplant exams based on practical experience. Mechanics who attain these top ratings have an increased opportunity to advance to higher paying jobs as lead mechanics, crew chiefs, inspectors, or shop foremen. Promotion to these higher grade jobs with the airlines is usually attained as a result of company seniority.

Applicants for a repairman certificate must have 18 months of practical experience in the maintenance duties of the specific job for which the person is to be employed by the repair station or have completed formal training acceptable to FAA. Avionic repair stations employ technicians who might be required to hold an FCC license.

Mechanics with advanced ratings and administrative ability can reach supervisory and executive positions, while those who have broad experience in maintenance and overhaul facilities can become designated inspectors for the FAA. Mechanics with the necessary pilot licenses and flying experience may take FAA examinations for the position of flight engineer, with eventual opportunities to become pilots.

Requirements to Enter the Job. While a high school diploma is not required to become an apprentice aircraft mechanic, employers give preference to applicants who are high school or vocational school graduates; thus, such a diploma is essential. Mathematics, physics, computer science, chemistry, English, and aerospace education courses are suitable subjects to pursue while in high school, because the aircraft mechanic/avionics technician must understand the physical principles involved in the operation of the aircraft and its systems. Also, a high school diploma is normally recommended as a prerequisite for attending a technical school or a college offering A&P training. The aircraft mechanic is expected to continue his or her education, even after hiring, in order to keep abreast of the continuing technical changes and improvements in aircraft and associated systems.

The successful aircraft mechanic should have an above average mechanical ability and a desire to work with his or her hands. He or she should also have an interest in aviation, appreciation of the importance of doing a job carefully and thoroughly, and the desire to learn throughout a career.

Opportunities for Training. As a qualified student who wants to become an aircraft mechanic you can follow one of several paths:

- Begin work for an airline or an independent repair station as an apprentice mechanic, learning as you earn. This method of earning an A&P/Repairman's Certificate or the FCC license normally takes longer and earning power remains at a lower rate over a longer period of time.
- Take aircraft mechanic courses at one of the many FAA certificated private or public technical schools. A high school diploma is normally recommended for entrance to these schools, but the period of training is normally shorter than on-the-job training and earnings upon completion of the course are higher. Also, as a graduate of such a course, you are then qualified to take the FAA exams.
- Receive training as an aircraft mechanic while in the military service, and with some additional study, you can qualify for a civilian mechanic job when the period of military service is completed.

Public and private vocational institutions along with the military services are major suppliers of aviation mechanics. In the past, many airlines had standing orders with FAA-approved schools and other educational institutions for all graduate mechanics, but the recent recession combined with deregulation of the airline industry has decreased job opportunities. At the present time, many licensed A&P mechanics are not working in their chosen field. The price of technical school training is expensive, costing several thousand dollars for an 18- to 24-month course. Fortunately, financial assistance is available through the U.S. Department of Education. For information, write to:

OFFICE OF STUDENT FINANCIAL ASSISTANCE
400 Maryland Ave. S.W.
Washington, D.C. 20202

A free list of FAA-certificated aviation maintenance technician schools (Advisory Circular 147-2W) is available from

U.S. DEPARTMENT OF TRANSPORTATION
Publication Section, M-494.3
Washington, D.C. 20590

The World Aviation Directory, which is available in the reference section of many libraries, has the most comprehensive listing of aircraft operators, manufacturers, and associated companies that design, produce, overhaul, and maintain aircraft. Finally, the following airline associations can furnish a list of their members:

AIR TRANSPORT ASSOCIATION OF AMERICA
1709 New York Ave. N.W.
Washington, D.C. 20006

REGIONAL AIRLINE ASSOCIATION
1101 Connecticut Ave. N.W.
Suite 700
Washington, D.C. 20036

Outlook for the Future. The long-term employment outlook for aviation maintenance personnel is very encouraging. According to the FAA, over the next few years there will be an annual average of 10,000 job openings for aircraft avionics maintenance personnel, increasing to 40,000 openings per year by 1990. These numbers are the result of analysis of anticipated aviation industry growth rates and projected retirements of the World War II and Korea era veterans, who presently hold many of the aviation maintenance jobs. Other studies are less optimistic about employment opportunities, but all emphasize the fact that the well-trained, licensed individual with a strong background in technical subjects will have little trouble finding work in aviation or associated technical fields. According to the Future Aviation Professionals of America (FAPA), for each new pilot hired, airlines will hire an additional mechanic. The cumulative ten-year demand predicted by FAPA shows that by 1997, over 40,000 more mechanics will be needed to keep up with the expanding airline market.

INSTRUMENT TECHNICIAN

The aviation instrument technician installs, tests, repairs, and overhauls all aircraft engine and navigational instruments. Almost all the indicators, meters, gauges, and other panel-mounted instruments used by the pilots would be maintained by the instrument technicians. These instruments include pitot static instruments such as altimeters, airspeed indicators, tachometers, fuel flow gauges, temperature indicators, and pressure gauges.

Usually, the flight line instrument technician is the one who removes the faulty unit from the aircraft, and the shop instrument technician overhauls the faulty unit in a sterile, carefully controlled shop atmosphere. Flight line technicians must usually hold a valid FCC license, but, the shop technician usually does not.

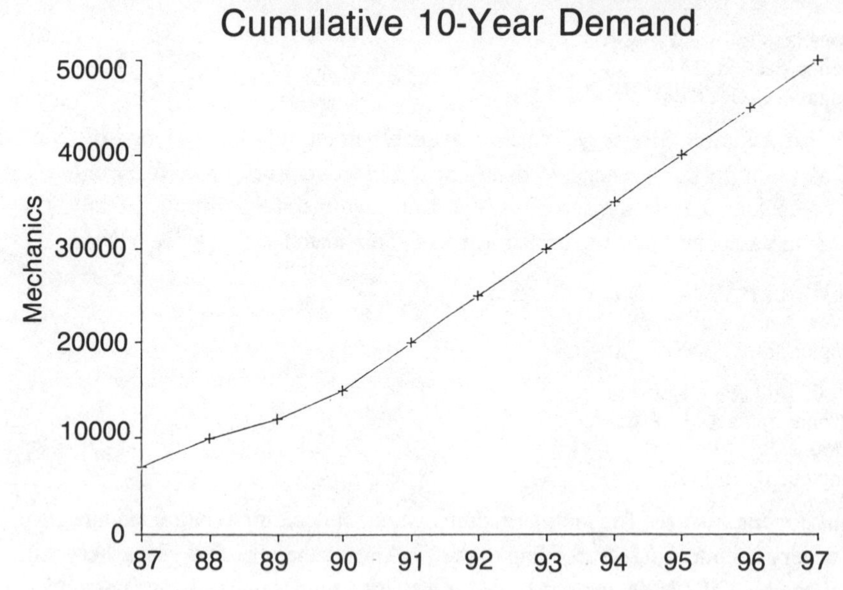

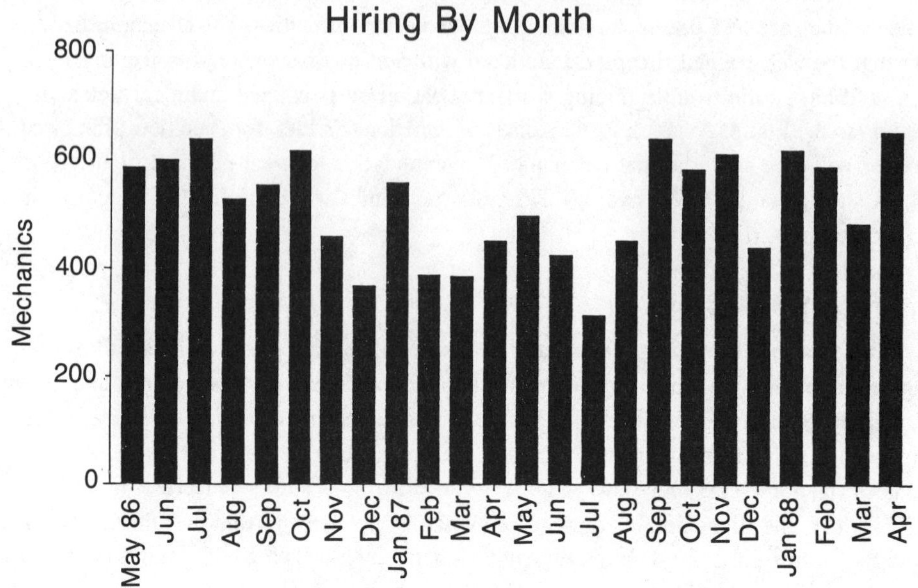

FAPA, Atlanta, GA 1-800-JET-JOBS

RADIO TECHNICIAN

Radio technicians install, test, repair, and overhaul all aircraft radio equipment. This includes navigation, communication and identification systems, such as two-way voice communication systems, VOR navigation, Doppler navigation (DNS), inertial navigation (INS), R-NAV navigation, omega navigation, automatic direction finding equipment (ADF), distance-measuring equipment (DME), instrument landing system (ILS), transponder systems (IFF), weather radar systems, automatic pilot systems, and all interfacing wiring and circuitry.

Two-way radio systems provide the flight crew with the ability to communicate with the various ground control authorities during flight. Back-up radios are installed in case of inflight radio problems. The navigation systems provide the pilot with continuous bearing, position, and desired track information. The instrument landing system provides the flight crew with glideslope, localizer, and marker beacon information during the approach to landing phase of the flight. The transponder system provides identification and position information to the ground-based air traffic controllers. The autopilot system helps to reduce the pilot workload by electronically controlling the flying characteristics of the aircraft during cruise. The weather radar system provides a warning of inclement weather along the desired route. This advanced warning allows the pilot time to make any necessary changes or deviations from the filed flight plan.

This electronic radio "package" is the heart of the modern-day airliner, airline travel, as we know it, could not exist without the proper repair, calibration, and maintenance provided by the radio maintenance technician.

Most aviation maintenance facilities are divided into two groups of technicians: the flight line technicians and the shop technicians. Flight line technicians perform "on aircraft" repairs by removing and replacing faulty units and repairing faulty interface circuitry. "Shop technicians" repair the faulty units removed from the aircraft in a well-equipped shop atmosphere.

Career Profile

Aviation Maintenance

Name:		G.D. Smith
Career Field:		Aviation Maintenance
Position Held:		Director of Maintenance Southwest Airlines
Education:	**1968**	Airframe and Powerplant Mechanic License Federal Aviation Administration
Experience		
1977 - Present:		*Director of Maintenance* Southwest Airlines, Dallas, Texas Responsible for all maintenance performed on Southwest Airlines aircraft.
1971 - 1977:		*Maintenance Superintendent*; Braniff Airlines, Dallas, Texas Supervised check and overhaul area for base maintenance.
1968 - 1971:		*General Foreman;* Braniff Airlines, Dallas, Texas General supervision responsibilities.
1959 - 1968:		*Aviation Maintenance Instructor*; Braniff Airlines; Dallas, Texas Instructed pilots on the theory and operation of various aircraft systems.
1958 - 1959:		*Electrical Foreman;* Braniff Airlines; Dallas, Texas General supervision responsibilities.
1952 - 1958:		*Electrical Mechanic;* Braniff Airlines; Dallas, Texas Performed electrical maintenance on various Braniff aircraft.
1951 - 1952:		*Apprentice Electrician;* Braniff Airlines; Dallas, Texas Performed general maintenance and repair as an apprentice electrician.

Job Responsibilities

As Director of Maintenance for Southwest Airlines, I am responsible for all maintenance performed on the Southwest fleet. The safety of aircraft operated in the National Airspace System is dependent in a large degree upon the satisfactory performance of all aircraft systems.

Reliability and performance of the system(s) is proportional to the quality of maintenance received and the knowledge of those who perform such maintenance. It is, therefore, important that maintenance be accomplished using the best techniques and practices to assure optimum performance.

Every time an aviation mechanic performs a job and signs the aircraft forms off, he or she is assuming a great deal of responsibility for the work accomplished. To do the job efficiently, the mechanic must be properly trained and have the background education necessary to perform the job.

Systems that require maintenance include propulsion, airframe, radio and electronic, hydraulic, pneumatic, environmental control, fuel, landing gear, flight control, and other various systems. Usually, aircraft mechanics perform a variety of maintenance tasks. You might be changing tires one day and engines the next. It all depends on the schedule and the workload.

Aviation mechanics perform two different types of maintenance: preventative and corrective. *Preventative* maintenance includes scheduled inspections and routine servicing. These measures are taken to help prevent malfunctions from occurring on the aircraft at a very inopportune time. *Corrective* maintenance is performed once a malfunction has been identified on the aircraft and must be corrected in order to bring the aircraft back into service.

To repair the aircraft, maintenance mechanics troubleshoot to determine the source of the problem, and remove, repair and replace the faulty unit(s). After the repair is accomplished, an operational check is accomplished to verify the repair.

Benefits and Drawbacks of the Job

Working as an airline aircraft mechanic has many definite benefits. Perhaps the most rewarding benefit is the job satisfaction you feel after a hard day's work. Unlike many other jobs, the mechanics field allows the worker to see a job through from beginning to end.

Compared to mechanics in other industries, aviation mechanics receive excellent compensation. Starting wages for mechanics are very competitive. With a few years experience, the motivated individual can move up into the supervisory and foreman ranks if he has the proper attitude, experience, and education.

As a mechanic, some travel might also be necessary. Not only do mechanics work out of their home base, but if an aircraft is broken out in the field, they might be expected to travel to the aircraft and repair the faulty system. These trips could last for hours or days, depending on the complexity of the repair and the distance of travel. While on business, all authorized expenses incurred on the trip are paid for by the airline.

Airline employees also enjoy unlimited travel benefits. Employees fly for free on their parent airline and at reduced rates on other participating airlines. If you love to travel, this could be a big plus.

Because most airlines fly almost 18 hours a day, seven days a week, the maintenance aspect is an ongoing concern. To keep the fleet flying, the maintenance crew typically works 24 hours a day, seven days a week. Also, due to the fact that most flights are during the day, much of the necessary aircraft maintenance is done at night. This type of schedule requires shift work, so you might be required to work rotating schedules. This could be a drawback if you are interested in a 9-to-5 job or a benefit if you enjoy the variety.

Also a consideration is that this type of job requires weekend and holiday work. Typically, airlines observe some of their busiest schedules during holidays, so instead of enjoying turkey dinner at home with Mom and Dad at Thanksgiving, you might be repairing a faulty inertial navigation system on an aircraft scheduled to fly in a few hours.

Many times when working on the flight line, aviation maintenance mechanics work in a noise-hazard environment. To prevent any adverse effects, earplugs, headsets, or other forms of hearing protection are frequently utilized or required. Another aspect of flight line work is that of the environment. During beautiful, summer days, working in the fresh air and sunshine might be a rewarding experience, but during times of inclement weather, the work must also be done. In the summer, temperatures on the flight line in many parts of the country can exceed 110°F during the peak hours of the day. Rain, snow, ice, and humidity are also factors that could make the job uncomfortable.

In lieu of working on the flight line, some mechanics work in a designated repair shop on equipment removed for repair. The shop atmosphere is usually climate controlled, and the mechanic can enjoy many of the comforts of home. Other benefits enjoyed by airline employees include generous health, retirement, and medical programs.

Recommendations to Job Hunters

Working as an aviation mechanic requires an individual who is substantially mechanically inclined and is not afraid to work hard. At Southwest, and at most airlines, personnel looks for applicants who have an Airframe and Powerplant (A&P) License and some practical experience.

It is important that the applicant is a high school graduate. College is not a necessity for the job; however, it could be helpful for the individual a few years down the road if he or she is interested in pursuing a supervisory or management position. There are many fine trade schools, such as Spartan School of Aeronautics in Tulsa, OK, which can train the individual for the Airframe and Powerplant License. See Chapter 2 for a listing of the FAA-certificated maintenance schools.

After earning the A&P license, the next step is to seek out an airline or fixed-base operator (FBO) that will hire you without the practical experience. Many of the other smaller organizations only hire applicants with at least two years of practical experience. Where to get the experience can be a problem for the recent A&P graduate (and any other graduate for that matter). Many larger organizations are willing to train you on their equipment and will allow you the training period; others will not. Consult with each potential employer and determine their specific requirements.

Another way to gain practical experience is in the military. Many military programs not only send the individual to an excellent training school, but they also provide a few years of on-the-job training. This type of experience is an excellent asset to the potential employee. My only caution here is try not to become too specialized. We like to see employees who are well rounded and understand many different types of aircraft systems. An applicant who is only experienced at changing tires would not be of much help in a diversified maintenance shop. Some of the larger airlines such as American and Delta do hire more specialized people because of various contract stipulations and requirements. But for the most part, well-roundedness is the general rule.

My last comment about pursuing such a position concerns the interview process. During the interview, I like to see confident and energetic individuals. The "I'm the mechanic you've been looking for" attitude really makes a lasting impression, and goes a long way to help you get the job.

AIRLINE ADMINISTRATION AND SALES

The airlines would not be able to function without those who supply the income (sales to customers) and those who manage the income (administration).

ADMINISTRATIVE POSITIONS

The airlines administrative network is the backbone of the organization and helps to direct the day-to-day operations of the airline. Many professionals are required to oversee the administrative, financial, and marketing aspects of the organization. These personnel supply upper management with the data necessary to plan future operations.

CLERICAL SUPPORT

There are many clerical and office positions classified within the airline industry. Typically clerical, support specialists perform such duties as bookkeeping, typing, and maintenance of records and files. They also transcribe dictation, answer phones, and handle mail and other forms of correspondence while utilizing a variety of business machines when necessary.

The successful applicant, depending on the position, should have anywhere from one year of experience and the ability to type to a college degree in business. Position is usually commensurate with education and experience.

CUSTOMER SERVICE AGENT

Customer service agents provide an interface between the airline and its customers. The two most common positions are Reservation Agents and Ticket Agents.

Reservation agents provide travel information over the telephone to customers. Such information includes the availability of space on a flight, trip planning, car rentals, hotel accommodations, information concerning fares, schedules, routes, tours, meals, and other pertinent information relative to the customer's travel plans. Reservation agents can also make the reservations necessary over the telephone. Ticket agents typically work at the airport behind the baggage counter. These agents sell tickets to passengers, check baggage, and provide necessary information on aircraft boarding. The successful customer service agent applicant should be a high school graduate or equivalent and have a neat appearance.

Table 4-1 is a list of U.S. major, national, and regional air carriers. Correspondence for employment should be addressed to the personnel department of the airline(s) of interest. For a more expanded listing of airlines and cargo carriers, refer to the *World Aviation Directory*, available in the reference section of most libraries.

Table 4-1. Major, National, and Regional Airlines

Major Airlines

American Airlines
P.O. Box 619616
Dallas/Ft. Worth Airport,
TX 75261-9616
(817)355-1234

Continental Airlines
P.O. Box 4607
Houston, TX 77210-4607
(713)630-5000

Delta Airlines Hartsfield Atlanta International Airport
Atlanta, GA 30320-9998
(404)765-2600

Eastern Airlines
Miami International Airport
Miami, FL 33148-0001
(305)873-2211

Northwest Airlines Minneapolis/St. Paul International Airport
St. Paul, MN 55111-3075
(612)726-2111

Pan-American World Airways
Pan American Building
200 Park Ave.
New York, NY 10166-0001
(212)880-1234

Piedmont Airlines
One Piedmont Plaza
Winston Salem, NC 27156-1000
(919)770-8000

Trans World Airlines
605 Third Ave.
New York, NY 10158-9616
(212)692-3000

United Airlines
P.O. Box 66100
Chicago, IL 60666-0100
(312)952-4000

U.S. Air
Washington National Airport
Washington, D.C. 20001-4997
(703)892-7000

Pacific Southwest Airlines
3225 N. Harbor Dr.
San Diego, CA 92101
(619)574-2100

Southwest Airlines, Inc.
Box 37611, Love Field
Dallas, TX 75235
(214)353-6100

World Airways
P.O. Box 2330
Oakland International Airport
Oakland, CA 94614-0330
(415)577-2000

National Airlines

Air Wisconsin
Outagamie Airport
Appleton, WI 54915
(414)739-5123

Alaska Airlines, Inc.
Box 68900
Seattle, WA 98168
(206)433-3200

Aloha Airlines, Inc.
P.O. Box 30028
Honolulu, HI 96820
(808)836-4101

American Trans Air, Inc.
Box 51609
Indianapolis International Airport
Indianapolis, IN 46251
(317)247-4000

American West Airlines, Inc.
222 S. Mill Ave.
Tempe, AZ 85281-2869
(602)894-0800

ARROW AIR, INC.
7955 NW 12th St.
Miami, FL 33126
(305)594-8080

BRANIFF, INC.
P.O. Box 7035
7701 Lemmon Ave.
Dallas, TX 75209
(214)358-6011

DHL AIRWAYS, INC.
333 Twin Dolphin Dr.
Redwood City, CA 94065
(415)593-7474

EVERGREEN INTERNATIONAL AIRLINES, INC.
3850 Three Mile Lane
McMinnville, OR 97128-9409
(503)472-0011

HAWAIIAN AIRLINES, INC.
P.O. Box 30008
Honolulu, HI 96820
(808)525-5511

MIDWAY AIRLINES
5700 S. Cicero
Chicago, IL 60638
(312)838-0001

REGIONAL AIRLINES

AERO VIRGIN ISLANDS CORP.
Box 546, Cyril E. King Airport
St. Thomas, VI 00801
(809)776-7725

AIR KENTUCKY AIR LINES
P.O. Box 797
Clifton Park, NY 12065
(518)383-3650

AIR MIDWEST, INC.
P.O. Box 7724
Wichita, KS 67277
(316)942-8137

AIR NEVADA
Box 11105
Las Vegas, NV 89111
(702)736-2702

AIR NEW ORLEANS, INC.
2nd Floor, 3545 N. 110 Service Rd.
Metairie, LA 70002
(504)887-1996

AIR RESORTS AIRLINES
2192 Palomar Airport Rd.
Carlsbad, CA 92008
(619)438-3600

ALLIANCE AIRLINES
9420 W. 52nd St.
Kenosha, WI 53412
(414)658-2025

AMERICAN EAGLE/WINGS WEST AIRLINES
P.O. Box 8115
San Luis Obispo, CA 93403-8115
(805)541-1100

ATLANTIC SOUTHEAST AIRLINES, INC.
1688 Phoenix Pkwy.
College Park, GA 30349
(404)996-4562

AVAIR, INC.
The Forum, 8601 Six Forks Rd.
Raleigh, NC 27615
(919)848-2700

BAR HARBOR AIRLINES
Bangor International Airport
138 Main Ave.
Bangor, ME 04401
(207)989-1090

BROCKWAY AIR
1 Calkins Ct.
S. Burlington, VT 05401
(802)658-5071

CATSKILL AIRWAYS, INC.
Box 72 I, RD-4
Oneonta, NY 13820
(607)432-8222

CHALKS INTERNATIONAL AIRLINE, INC.
1550 SW 43rd St.
Ft. Lauderdale, FL 33315
(305)947-1308

CHAPARREL AIRLINES
P.O. Box 206
Abilene, TX 79604
(915)677-3337

CHAUTAUQUA AIRLINES, INC.
P.O. Box 505
Chautauqua Airport
Jamestown, NY 14702
(716)664-2400

CHRISTMAN AIR SYSTEM
P.O. Box 776
Washington, PA 15301
(412)228-1600

COLGAN AIRWAYS CORP.
Box 1650
Municipal Airport
Manassas, VA 22110
(703)631-2230

COMAIR, INC.
P.O. Box 75021 Airport Dr.
Greater Cincinnati Airport
Cincinnati, OH 45275
(606)525-2550

COMMAND AIRWAYS
Dutchess County Airport
Wappingers Falls, NY 12590
(914)462-6100

CORPORATE AIR, INC.
Bldg. 85-214
Bradley International Airport
Windsor Locks, CT 06096
(203)623-9119

CROWN AIRWAYS, INC.
P.O. Box 377
DuBois-Jefferson County Airport
Falls Creek, PA 15840
(814)371-2691

CUMBERLAND AIRLANES
Box 1611
Cumberland, MD 21502
(304)738-8640

EAST HAMPTON AIRE, INC.
P.O. Box CCC
East Hampton, NY 11937
(516)537-0560

GP EXPRESS AIRLINES, INC.
P.O. Box 218
Central Nebraska Regional Airport
Grand Island, NE 68802-0218
(907)822-5532

GREAT LAKES AVIATION, LTD.
Box 115A, R.R.-3
Spencer Municipal Airport
Spencer, IA 51301
(712)262-7734

HENSON AIRLINES
Salisbury-Wicomico County
Regional Airport
Salisbury, MD 21801
(301)742-2996

HOLIDAY AIRLINES, INC.
Terminal A
Newark International Airport
Newark, NJ 07114
(201)961-2770

IOWA AIRWAYS
Municipal Airport
Dubuque, IA 52001
(800)526-4692

JETSTREAM INTERNATIONAL AIRLINES
Suite 400, 6520 Poe Ave.
Dayton, OH 45414
(513)454-1116

LONG ISLAND AIRLINES
Gate 2, Republic Airport
Farmingdale, NY 11735
(516)752-8300

MALL AIRWAYS, INC.
Albany County Airport
Albany, NY 12211
(518)785-7525

MESABA AVIATION, INC.
6201 34th Ave. S
Minneapolis, MN 55450
(612)726-5151

METRO AIRLINES, INC.
P.O. Box 612626
DFW Airport, TX 75261
(214)929-3400

MIDSTATE AIRLINES, INC.
3101 Dixon St.
Stevens Point, WI 54481
(715)346-8000

MIDWAY COMMUTER
Capital City Airport
Springfield, IL 62708
(217)789-0091

NPA, INC.
Bldg. 142
Tri-Cities Airport
Pasco, WA 99301
(509)545-6420

NASHVILLE EAGLE, INC.
Metropolitan Airport
Nashville, TN 37217
(615)275-3753

NEW ENGLAND AIRLINES, INC.
State Airport
Westerly, RI 02891
(401)596-2460

PAN AM EXPRESS
N.E. Philadelphia Airport
Grant Ave. & Ashton Rd.
Philadelphia, PA 19114
(215)934-6111

PROVIDENCETOWN BOSTON AIRLINE, INC. (PBA)
3201 Radio Rd.
Naples, FL 33942
(813)774-8822

PENNSYLVANIA AIRLINES, INC.
Harrisburg International Airport
Middletown, PA 17057
(717)944-2781

POCONO AIRLINES, INC.
Wilkes-Barre/Scranton International Airport
Avoca, PA 18641
(717)655-2989

PRECISION VALLEY AVIATION
Manchester Municipal Airport
841 Galaxy Way
Manchester, NH 03105
(603)668-0082

PRINCEVILLE AIRWAYS, INC.
Honolulu International Airport
Commuter Airline Terminal
Honolulu, HI 96819
(808)833-3219

ROCKY MOUNTAIN AIRWAYS, INC.
Hangar 6
Stapleton International Airport
Denver, CO 80207
(303)388-8585

ROYALE AIRLINES, INC.
6129 Amelia Earhart Dr.
Shreveport, LA 71109-7797
(318)635-8168

SAN JUAN AIRLINES
1402 Fairchild International Airport
Port Angeles, WA 98362
(206)452-5326

SCENIC AIRLINES, INC.
241 E. Reno Ave.
Las Vegas, NV 89119
(702)739-5611

SIMMONDS AIRLINES, INC.
Marquette County Airport
Negaunee, MI 49866
(906)475-9901

SKYWEST AIRLINES
Suite 201, 50 E, 100 S
St. George, UT 84770
(801)628-2655

SOUTHERN JERSEY AIRWAYS, INC.
Bader Field Airport
Atlantic City, NJ 08401
(609)348-4600

STATESWEST AIRLINES
Sky Harbor International Airport
2871 Sky Harbor Blvd.
Phoenix, AZ 85094
(602)220-0391

SUBURBAN AIRLINES, INC.
P.O. Box 1201
Reading Municipal Airport
Reading, PA 19603
(215)375-8551

TENNESSEE AIRWAYS, INC.
McGhee Tyson Airport
Alcoa, TN 37701
(615)970-3100

VIRGIN AIR, INC.
P.O. Box 2788
St. Thomas, VI 00801
(809)776-2722

WESTAIR COMMUTER AIRLINES, INC.
5570 Air Terminal Dr.
Fresno, CA 93727
(209)294-6915

WHEELER FLYING SERVICE
Box 12034
Research Triangle Park, NC 27709
(919)787-8833

WINGS AIRWAY
Wings Field
Stenton Ave. and Narcissa Rd.
Blue Bell, PA 19422
(215)646-1800

WISE AVIATION
Municipal Airport
Clinton, IA 52732
(319)242-5721

5

Careers with the Government

T-6 Texan

A major source of aviation careers lies in jobs with federal, state, and local government agencies. The following information is courtesy of the Aviation Career Services pamphlet GA-300-128-84.

Civil aviation careers in the federal government for men and women are found within the Department of Transportation, the Federal Aviation Administration, and a growing number of other federal departments and agencies. All of these aviation jobs come under the Federal Civil Service, and wage scales are determined by Congress, which, from time to time adjusts the pay levels to bring them in line with comparable jobs in private business and industry. Salaries for Federal Civil Service employees are established into two chief categories: General Schedule (for those employees who perform administrative, managerial, technical, clerical, and professional jobs and who are paid on an annual basis) and the Federal Wage System (for those employees who perform jobs associated with the trades and crafts and who are paid wages on an hourly basis).

Most Federal Civil Service employees in the aviation field are covered by the General Schedule and their salaries vary according to their grade level (GS-1 through GS-18). See Table 5-1.

Table 5-1. 1989 FEDERAL PAY SCHEDULE

	YEARS									
	1	2	3	4	5	6	7	8	9	10
GS-1	$10,213	$10,555	$10,894	$11,233	$11,573	$11,773	$12,108	$12,445	$12,461	$12,780
2	11,484	11,757	12,137	12,461	12,601	12,972	13,343	13,714	14,085	14,456
3	12,531	12,949	13,367	13,785	14,203	14,621	15,039	15,457	15,875	16,293
4	14,067	14,536	15,005	15,474	15,943	16,412	16,881	17,350	17,819	18,288
5	15,738	16,263	16,788	17,313	17,838	18,363	18,888	19,413	19,938	20,463
6	17,542	18,127	18,712	19,297	19,882	20,467	21,052	21,637	22,222	22,807
7	19,493	20,143	20,793	21,443	22,093	22,743	23,393	24,043	24,693	25,343
8	21,590	22,310	23,030	23,750	24,470	25,190	25,910	26,630	27,350	28,070
9	23,846	24,641	25,436	26,231	27,026	27,821	28,616	29,411	30,206	31,001
10	26,261	27,136	28,011	28,886	29,761	30,636	31,511	32,386	33,261	34,136
11	28,852	29,814	30,776	31,738	32,700	33,662	34,624	35,586	36,548	37,510
12	34,580	35,733	36,886	38,039	39,192	40,345	41,498	42,651	43,804	44,957
13	41,121	42,492	43,863	45,234	46,605	47,976	49,347	50,718	52,089	53,460
14	48,592	50,212	51,832	53,452	55,072	56,692	58,312	59,932	61,552	63,172
15	57,158	59,063	60,968	62,873	64,778	66,683	68,588	70,493	72,398	74,303
16	67,038	69,273	71,508	73,743	75,473	76,678*	78,869*	81,060*	82,500*	
17	76,990*	79,556*	82,122*	82,500*	83,818*					
18	86,682*									

* The rate of basic pay payable to employees at these rates is limited to the rate for level V of the Executive Schedule, which would be $75,500.

SENIOR EXECUTIVE SERVICE

ES-1	$68,700
ES-2	71,800
ES-3	74,900
ES-4	76,400
ES-5	78,600
ES-6	80,700

EXECUTIVE SCHEDULE

level I	$99,500
level II	89,500
level III	82,500
level IV	80,700
level V	75,500

EXPERIENCE REQUIREMENTS

The following table shows an example of the requirements for a particular government job. The terms and substitutions are explained subsequently.

	Experience in Years		
Grade	General	Specialized	Total
GS-9	3	2	5
GS-11/15	3	3	6

General Experience. General experience is that which has provided familiarity with aircraft operation or the aviation industry. The following are examples of qualifying general experience:

- Pilot or crew member in civil or military aviation
- Civilian or military air traffic controller
- Aviation mechanic or repairperson
- Avionics or electronics technician
- Skilled machinist, assemblyperson, or inspector in production of aircraft, aircraft parts, or avionics equipment.

Substitution of Education. For all positions in this series, successful completion of post-high-school education in related fields such as engineering, aeronautics, or air transportation can be substituted for the required general experience. Education may be substituted at the rate of one academic year of full time study for nine months of general experience, up to the maximum of three years of general experience.

Specialized Experience. Specialized experience is that which has provided knowledge and skills for work in the specialty field operations, airworthiness, or manufacturing. In addition, specialized experience must have provided a broad knowledge of the aviation industry, the general principles of aviation safety, and the federal laws, regulations and policies regulating aviation. Examples of qualifying specialized experience are described under the appropriate specialty area.

Level of Experience. Candidates for positions at grades GS-11 and below must have had at least six months of specialized experience at a level of difficulty and responsibility comparable to that of the next lower grade in the federal service or one year equivalent to the second lower grade. Candidates for grades GS-12 and above must have had at least one year of specialized experience of difficulty and responsibility comparable to that of the next lower grade.

For any grade, the required amount of experience and education is not in itself accepted as proof of qualification for a position. The candidate's total record of experience and education must demonstrate that he or she possesses the ability to perform the duties of the position.

Within each of the grades provided in the General Schedule, provision is made for periodic pay increases based on an acceptable level of performance. With an acceptable level of competence, the waiting period of advancement to steps two, three, and four is one year, steps five, six, and seven is two years, and steps eight, nine, and ten is three years.

Forty hours constitutes a normal work week. Additional payment (called premium pay) is made for shift work involving duty between 6:00 p.m and 6:00 a.m. and for work during Sundays and holidays. Merit promotions are awarded under provisions of a Civil Service approved merit promotion plan.

Most federal employees under Civil Service participate in a liberal retirement plan. Employees earn from 13 to 26 days of paid annual vacation, depending upon the length of service, and 13 days of paid sick leave each year. Health insurance, low-cost group life insurance, and credit union service are other benefits offered.

To apply for a government job, application must be made on form SF-171 (Personal Qualifications Statement) for all federal jobs. Forms can be obtained from your nearest Office of Personnel Management/Federal Job Information Center (see Table 5-2 for a complete listing).

Table 5-2. STATE AVIATION JOBS

JOB	NATURE OF THE WORK
Director	Promotes aviation in the state, administers state aviation regulations, represents the state at regional meetings, and directs the staff of the Department of Aeronautics.
Deputy and Assistant Director	Assists the Director.
Administrative Assistant	Handles the detailed routine operation of the Director's office.
Pilots	Fly state-owned aircraft as required, for example to take the Governor to meetings. (*Note:* Some departments who do not employ pilots require the Director, his assistant, or some other staff member to assume pilot duties when necessary).
Field Service Representative	Is in direct contact with aviation interests within the state. Might be called upon to explain proposed or new flying rules, help with aircraft sound problems, or assist with an aviation education project.
Accountants and Statisticians	Maintain financial records of the department and gather flight statistics about aircraft movements, registered pilots, accidents, hours flown, etc.
Stenographers, Clerks, and Typists	Carry on routine office duties.
Engineers	Civil, electronics, radio, and other engineering specialities involved in planning airports and improvements to airports, installing and supervising air navigational aids operated by the state.
Chief Planner	Responsible for preparation of State's airport system plan and other planning activities.

Engineer Technicians (Aides)	Assist engineers in their work (draftsperson, etc.).
Aeronautical Inspectors	Check compliance with state aviation regulations.
Aviation Education Officers	Carry out aviation education policies of the Department. Cooperate with schools in aerospace education programs.
Publication Editors	Are responsible for publication of newsletters, releases, and other information of interest to pilots, airport operators, and fixed base operators in the state.
Safety Officers	Promote aviation safety, such as conducting weather seminars and other safety-type meetings for pilots.
Aircraft Mechanics	Service and maintain state-owned aircraft.
Surplus Property	Search out surplus federal government property that might be useful to state aviation.

FEDERAL AVIATION ADMINISTRATION (FAA)

The largest number of aviation jobs found within the federal government (outside the Department of Defense) is with the FAA, which is a division of the Department of Transportation. The FAA, with a total of approximately 47,000 employees, is charged with the administration and enforcement of all federal air regulations to ensure the safety of air transportation. The FAA also promotes, guides, and assists the development of a national system of civil airports. The FAA provides pilots with flight information and air traffic control services from flight planning to landing.

The following section lists and explains many of these FAA job titles, many of which are similar to those in the civilian sector.

AIR TRAFFIC CONTROL SPECIALIST (ATCS)

Requirements to Enter the Job. Applicants must have the following quantity (in years) and type of experience:

	Experience in Years		
Grade	General	Specialized	Total
GS-7	3	0	3
Entry Level	3	2	5
GS-9	3	3	6

General Experience. Progressively responsible experience in administrative, technical or other work which demonstrates potential for learning and performing air traffic control work. A four-year college degree can be substituted for the general experience requirement.

Specialized Experience. Experience in a military or civilian air traffic facility that demonstrates possession of the knowledge, skills, and abilities required to perform the level of work of the specialization for which application is made. Persons with a four-year college degree and a test score of 75.1 or higher on the Air Traffic Controller Aptitude Test will be admitted to controller training at the FAA Academy at Oklahoma City, OK.

Educational and Other Substitutions for Experience. Applicants considering the GS-7 level, for example, can use the following substitutions:

- ✦ Successful completion of four-year college degree can be substituted in full for the experience required at GS-7.
- ✦ Applicants who have passed the written test qualify for the experience requirements for grade GS-7 if they:
  - Hold or have held an appropriate facility rating and have actively controlled air traffic in civilian or military air traffic control terminals or centers;
  - Hold or have held an FAA certificate as a dispatcher for an air carrier;
  - Hold or have held an instrument pilot certificate;
  - Hold or have held an FAA certificate as a navigator or have been fully qualified as a navigator/bombardier in the Armed Forces;
  - Have 350 hours of flight time as a copilot or higher and hold or have held a private pilot certificate or equivalent Armed Forces rating;
  - Have served as a rated Aerospace Defense Command Intercept Director;
  - Meet the requirements for GS-5 and in addition pass the written test with a higher score.

Certificate and Rating Requirements. Air traffic control specialists in all specializations are required to possess or obtain a valid Air Traffic Control Specialist Certificate and/or Control Tower Operator Certificate, if appropriate. These certificates require demonstrating knowledge of basic meteorology, basic air navigation, standard air traffic control and communications procedures, the types and uses of aid to air navigation, and regulations governing air traffic. In addition, each air traffic control specialist must possess or obtain a rating for the facility assigned. This facility rating requires demonstration of a knowledge of the kind and location of radio aids to air navigation, the terrain, the landmarks, the communications systems and circuits, and the procedures peculiar to the area covered by the facility. All required certificates and ratings must be obtained, if not already held, within uniformly applicable time limits established by agency management.

Physical Requirements. Candidates must be able to pass a physical examination (including normal color vision). Air traffic control specialists are required to requalify in a physical examination given annually.

Written Test and Interview. Applicants must also pass a comprehensive written test and complete a personal interview during which alertness, decisiveness, diction, poise, and conciseness of speech are evaluated. Both men and women are employed as air traffic controllers. Few occupations make more rigid physical and mental demands upon employees than that of air traffic controllers. Because studies show that the unique skills necessary for success as a controller diminish with age, a maximum age of 30 has been established, without exception, for entry into an FAA tower or center controller position.

ATCS at FAA Airport Traffic Control Tower

Nature of the Work. The air traffic control specialists at FAA traffic control towers direct air traffic so it flows smoothly and efficiently. The controllers give pilots taxiing and takeoff instructions, air traffic clearances, and advice based on information received from the National Weather Service, air route traffic control centers, aircraft pilots, and other sources. They transfer control of aircraft on instrument flights to the Air Route Traffic Control Center (ARTCC) controller when the aircraft leaves their airspace and receives from the ARTCC control of aircraft on instrument flights flying in their airspace. They must be able to quickly recall registration numbers of aircraft under their control, the aircraft types and speeds, their positions in the air, and the location of navigational aids in the area.

Working Conditions. The controllers normally work a 40-hour week in FAA control towers at airports using radio, radar, electronic computers, telephone, traffic control lights, and other devices for communication. Shift work is necessary. Each controller is responsible, at separate times, for: giving taxiing instructions to aircraft on the ground, takeoff instructions and air traffic clearances, and directing landings of incoming planes. These individual duties are rotated among the staff about every two hours at busy locations. At busy times, controllers must work rapidly, and mental demands increase as traffic mounts, especially when poor flying conditions occur and traffic stacks up. Brief rest periods provide some relief, but are not always possible. Radar controllers usually work in semi-darkness.

Wages. The starting grade is normally GS-7. Trainees are paid while learning their jobs. The highest grade for a nonsupervisory professional air traffic control specialist in the tower is GS-14.

Opportunities for Advancement. Promotion from trainee to a higher grade professional controller depends on the employee's performance and satisfactory progression in his or her training program. Trainees who do not successfully complete their training courses are separated or reassigned from their controller positions. Increases in grade (with accompanying increases in salary) for successful trainees are fairly rapid, but grades above GS-14 are limited to managerial positions of team supervisor, assistant chief, staff officer and chief. During the first year, the trainee is on probation, then she or he can advance from positions backing up professional controllers to primary positions of responsibility. It takes a controller from three to six years of experience to reach the full performance level. Some professional controllers are selected for research activities with FAA's National Aviation Facilities Experimental Center in Atlantic City, NJ. Some are also selected to serve as instructors at the FAA Academy in Oklahoma City, OK.

Opportunities for Training. Trainees receive 15 weeks of instruction at the FAA Academy in Oklahoma City, OK. After completion of the training period, they are assigned to developmental positions for on-the-job training under close supervision until successful completion of training.

However, those who fail to complete training are separated or reassigned from their controller positions. The FAA conducts upgrading training programs for controllers continuously. Training in air traffic control continues long after the controller reaches the full performance level.

Outlook for the Future. Although aviation is growing dramatically, the number of new controllers hired each year is only approximately 2,000. This is due to the fact that advances in automation have allowed fewer controllers to do more work. There is, however, an increased emphasis on providing the maximum amount of safety that results in continued stringent requirements for controllers.

ATCS at FAA Air Route Traffic Control Center

Nature of the Work. The air traffic control specialists at FAA air route traffic control centers give pilots instructions, air traffic clearances, and advice regarding flight conditions along the flight path while the pilot is flying the federal airways or operating into airports without towers. The controllers use flight plans and keep track of progress of all instrument flights within the center's airspace. She or he transfers control of aircraft on instrument flights to the controller in the adjacent center when the aircraft enters that center's airspace. The controllers also receive control of flights entering his or her area of responsibility from adjacent centers. She or he monitors the time of each aircraft's arrival over navigation fixes and maintains records of flights under his or her control.

Working Conditions. Air route controllers work at FAA air route traffic control centers forty hours a week using electronic communications devices. Shift work is necessary. They work in semi-darkness, and unlike the tower controllers, never see the aircraft they control except as "targets" on the radarscope. In most areas, work is demanding. Registration numbers on all aircraft under control as well as types, speeds, and altitudes are automatically displayed on the radarscope, but each aircraft must be closely controlled to avoid other aircraft.

Where the Jobs Are. FAA employs about 6,800 controllers at 22 air route traffic control centers located throughout the U.S. plus one each in Guam and Puerto Rico.

Wages. The starting grade is normally GS-7. Trainees are paid while learning their jobs. The highest grade for an operating professional air traffic control specialist at a center is GS-14.

Opportunities for Advancement. Promotion to higher grades and to professional controller depends upon the employee's performance and satisfactory achievement in his or her training program. Increases in grade (with accompanying increases in salary) for successful trainees are fairly rapid, but grades above GS-14 are for positions of team supervisor, assistant chief, staff officer, and chief. During the first year, the trainee is on probation and he or she can advance from positions backing up professional controllers to primary positions of responsibility. It takes a controller from about three to six years of experience to reach the professional level. Appointment or movement to a position as air traffic control data systems computer specialist is possible. Professional controllers are also selected for research activities with FAA's National Aviation Facilities Experimental Center at Atlantic City, NJ. Some are also selected to serve as instructors at the FAA Academy in Oklahoma City, OK.

Opportunities for Training. Trainees receive 12 weeks of instruction at the FAA Academy in Oklahoma City, OK. After completion of the training period, they are assigned to developmental positions for on-the-job training under close supervision until successful completion of training. However, those who fail to complete training are separated or reassigned from their controller positions. The FAA conducts upgrading training programs for controllers continuously. Training in air traffic control continues long after the controller reaches the full performance level.

Outlook for the Future. In line with predictions for continued growth of all sections of aviation, the need for air traffic controllers will remain constant. As airports generate greater volumes of air traffic and as emphasis on providing the maximum amount of safety grows, there will be a continuing requirement for controllers at air route traffic control centers; however, automation has offset the increase in workload, thus eliminating the need for increases in the number of controllers.

ATCS AT FAA FLIGHT SERVICE STATION

Nature of the Work. The air traffic control specialists at FAA flight service stations render preflight, in-flight, and emergency assistance to all pilots on request. They give information about actual weather conditions and forecasts for airports and flight paths; relay air traffic control instructions between controllers and pilots; assist pilots in emergency situations; and initiate searches for missing or overdue aircraft.

Working Conditions. Shift work is necessary. They use a telephone, radio, teletypewriter, and direction finding and radar equipment. They work in office situations close to communications and computer equipment for 40 hours as a normal work week.

Where the Jobs Are. FAA flight service stations are found at approximately 317 locations throughout the United States, Virgin Islands, and Puerto Rico. About 4,300 flight service specialists are employed.

Wages. The starting grade is normally GS-7. Trainees are paid while learning their jobs. The highest grade for the flight service specialist is GS-11.

Opportunities for Training. Trainees receive 16 weeks of instruction at the FAA Academy in Oklahoma City, Oklahoma. After completion of the training period, they are assigned to developmental positions for on-the-job training under close supervision until successful completion of training. However, those who fail to complete training are separated or reassigned from their positions. The FAA conducts upgrading training programs for controllers continuously. Training in air traffic control continues long after the specialist reaches the full performance level.

Opportunities for Advancement. Excellent opportunities exist for the employee who successfully progresses in his or her training to attain higher grade levels as she or he gains experience and as the responsibilities and the complexity of duties increases. Beginning as a trainee in the flight service station, he or she can advance to an assistant chief and then to deputy chief or chief of the facility. As a further step, a few positions at higher grade levels are available in FAA regional offices with administrative responsibilities over all flight service stations within the area's jurisdiction.

Outlook for the Future. The number of specialists at flight service stations is not expected to increase as are jobs in other areas of air traffic control employment. Flight service stations will serve larger areas with the increasing use of long-distance telephone and other communications devices. Even though the number of opportunities for jobs for these specialists is not expected to increase greatly, these jobs will be more challenging as automation is introduced, and they will be stepping stones to air traffic controller careers in FAA-operated airport traffic control towers and at air route traffic control centers.

Career Profile

Air Traffic Controller

Name:		Oscar McNeil
Career Field:		Air Traffic Control (FAA)
Position Held:		Supervisory Air Traffic Controller ATC Facility; Dallas/Ft. Worth Airport; Dallas, Texas
Education:	**1970**	Air Traffic Control School, FAA Academy, Oklahoma City, OK
	1963	Bachelors Degree in Mathematics University of Texas at Austin
Experience:		
1987 - Present		Supervisory Air Traffic Controller Dallas/Fort Worth International Airport; Dallas, Texas
1986 - 1987		Air Traffic Controller Chicago O'Hare International Airport; Chicago, Illinois
1984 - 1986		Air Traffic Controller Intercontinental Airport; Houston, Texas
1977 - 1984		Air Traffic Controller Dallas/Fort Worth International Airport, Dallas, Texas
1970 - 1977		Air Traffic Controller San Antonio International Airport; San Antonio, Texas
1968 - 1970		Operations Specialist United States Army

Job Description

Nearly half of the work force of the Federal Aviation Administration (FAA) consists of people in the air traffic control field. Air Traffic Control Specialists (ATCSs) provide safety and informational services to the flying public. They provide radar, radio, weather, and other safety information services to pilots. ATCS work in three basic specialities: towers, enroute centers, and flight service stations (FSS).

In operating towers, ATCS control flights within the 3-to-30-mile radius each serves under visual flight rules (VFR) or instrument flight rules (IFR). Tower controllers work either in the glass-walled room at the top of the tower or in the radar room below it, but their jobs have the same aim—the safe separation and movement of planes within their area in taking off, landing, and maneuvering.

ATCSs who work in enroute centers control traffic that operates along established airways across the country between tower jurisdictions. They maintain a progressive check on aircraft, issuing instruction, clearances, and advice, as well as initiating search-and-rescue operations to locate overdue aircraft.

FSSs provide assistance to pilots who must obtain information on the station's particular area including terrain, weather peculiarities, preflight and in-flight weather information, suggested routes, altitudes, indications of turbulence, icing, and any other information important to the safety of a flight.

Recruitment Facts

The FAA has developed and adopted a mandatory centralized training program at the FAA Academy in Oklahoma City, Oklahoma, for potential air traffic controllers. To apply to the Academy, the interested applicant should submit a *CSC Form 5000B Admission Notice*, which are available at an area office of the United States Office of Personnel Management or by sending a postcard with your name and address to

SPECIAL EXAMINING DIVISION
P.O. Box 26650
ATTN: AAC-80
Oklahoma City, OK 73126

Application materials and testing information will then be mailed to you.

While attending the Academy Program in Oklahoma City, you will be paid a GS-7 salary. Because there are 26 pay periods a year, you will receive your paycheck every two weeks. To qualify for a GS-7, you must pass the written test plus possess three years of general experience, *or* four years of college, or any combination of education and experience equaling three years.

General experience is defined as progressively responsible work that demonstrates potential for learning and performing air traffic control work. This experience could have been gained in administrative, technical, or other types of employment. You must also be a United States citizen.

Many people ask if there is a study guide available for the new ATCS test. There is no OPM study guide available for the current edition of the ATCS test. In the past, private companies have developed study booklets; therefore an applicant can check with area bookstores. It is possible to pass the exam without prior ATCS or other aviation experience; however, a knowledge of aviation terms, procedures, and equipment increases the potential for a higher test score.

A maximum entry age of 30 years at the time of certification for appointment to FAA towers and centers has been established. The age limit of 30 does not apply to FSS positions.

Before you can be appointed, you must report for an interview and pass a complete physical examination by a medical examiner designated and paid for by the government. A background investigation is initiated for every applicant.

Initial training includes a demanding, 12-to-17-week course at the FAA Academy in Oklahoma City. Training costs, including travel from the Academy to the facility are paid by the agency. Trainees must progress through the developmental positions to full performance level in order to be retained by the FAA. Facility training is a combination of on-the-job training and classroom time that continues until the trainee reaches the journeyman level for the position assigned.

Air traffic facilities are located at many airport and major urban areas. Most FAA positions are filled on a regional basis. As an example, the Southwest Region includes Texas, New Mexico, Oklahoma, Louisiana, and Arkansas.

Successful candidates are referred to the regions for consideration in either flight service station (FSS) positions or center/tower positions. Initial placement decisions for FSS candidates will consider applicant's desires; however, assignments are primarily based on the agency's needs. Initial placement decisions for center/tower candidates are not made until they have completed the Academy screening course. Assignments will then be based on the agency's needs, Academy performance, and the employee's personal preference.

Current beginning salary is at the GS-7 level. Depending upon the air traffic control option, level of the facility, and staffing regulations, some journeyman grade levels can reach the GS-14 level.

Benefits and Drawbacks of the Job

Like every job, air traffic control work has its benefits and drawbacks. Air traffic control is a very demanding field of work because it takes people who can think clearly and quickly. It also requires confidence in yourself and your fellow controllers. During times of bad weather when flight schedules get backed up, the job can be extremely busy and somewhat pressuring. Depending on the person, some people handle pressure very well; in fact it helps them work more efficiently. On the other hand, those who cannot handle the pressure may find the work over-taxing and frustrating.

Another aspect of air traffic control work is that virtually all ATCS jobs involve shift work, because most facilities operate on a 24-hour basis. The exact rotation of the shift is usually determined by the individual facility. Days off will not always fall on weekends. Most employees at one time or another will work holidays and weekends. This type of schedule can be hard on family and social life. The FAA realizes that these hardships should be recognized, so you might receive additional compensation for the following: required work performed on holidays and Sundays, if your regular duty tour includes work between the hours of 6:00 p.m. and 6:00 a.m. (night differential pay), or if you work in excess of 8 hours in a day or 40 hours in the administrative work week (overtime pay).

Working for the government has its good points. The work can be very challenging and rewarding. Air traffic control is a very important job, and knowing that it has been done right can be a satisfying feeling at the end of a busy day. For the most part, working for the government has a certain amount of job security. There is a definite benefit working for one of the largest employers in the country.

Air traffic controllers also enjoy many other benefits. As an air traffic control specialist at the GS-9 level and above, you will not be considered available for active military service in time of war or national emergency. However, an exception might be recommended if your absence for reserve training or mobilization will not jeopardize the mission of the FAA facility to which you are assigned.

You earn annual leave for vacation and other purposes that require time away from your job according to the number of years (civilian plus creditable military service) you have been in federal service: 13 days a year for the first 3 years and 20 days a year for the next 12 years. After 15 years, you earn 26 days of annual leave each year. In addition to all of this, sick leave is earned at the rate of 13 days a year.

Comments/Recommendations to Job Applicants

Once you have the background experience necessary for the job, the next most important thing is to do well on the ATC exam. There are many fine schools that can help you to prepare for the FAA exam. They will teach you the background information necessary for the test. The military is also a good training ground for air traffic controllers. Whichever route you go, remember that there are age limitations on these jobs, so if you're interested, the time to get started is now.

Electronics Technician

Nature of the Work. The electronic technician installs and maintains electronic equipment required for aerial navigation, communications between aircraft and ground services, and control of aircraft movements to assure safety in the air and smoothly flowing air traffic. This involves work with radar, radio, computers, wire communications systems, and other electronic devices at airports and along the network of federal airways. It includes preventive maintenance (inspection of equipment, meter reading, replacement of deteriorating parts, adjustments) and corrective maintenance (troubleshooting, repair, and replacement of malfunctioning equipment). Electronic technicians may also specialize in design, development, and evaluation of new types of electronic equipment for the federal airways.

Working Conditions. They usually work out of an Airway Facilities Sector Field Office with other technicians whose work is directed by a supervisor. The office is frequently located at an airport and the equipment for which the office is responsible is within a 30- or 40-mile radius of the airport—in control towers, air route traffic control centers, flight service stations, or in open fields and even on remote mountain tops. Some of the work must be performed outdoors in all kinds of weather. Forty hours comprise a regular work week with shift work and weekend work rotated.

Where the Jobs Are. The FAA employs thousands of electronic technicians. Most electronic technicians work in field offices or "sectors" scattered all over the country. Some work is located at the FAA's National Aviation Facilities Experimental Center that is engaged in electronic research and development projects.

Wages. The entrance level normally starts at GS-5.

Opportunities for Advancement. The employee has opportunities to progress to higher grade levels depending upon the complexity of her or his duties, the degree of supervision received or exercised, and the growing knowledge and skills used in the performance of the work. Supervisory positions are available at sector, area, and regional offices. Promotion to managerial jobs at FAA Headquarters is possible.

Requirements for the Job. Age eighteen is the minimum age. Experience and education or training in electronics (a knowledge of basic electronic theory and related mathematics, transmitters and receivers, use of test equipment, techniques of troubleshooting and circuitry analysis, use of tools, and installation practices) are required. The greater the degree of education and/or experience, the higher the entrance level. Applicants must have had a minimum experience of the kinds and amounts indicated in the table below for each grade. Excess "specialized" experience may be credited as "general" experience. Some types of civilian or military education related to the option for which application is made can be substituted for the specialized experience requirement.

	Experience in Years		
Grade	General	Specialized	Total
GS-5	3	0	3
GS-6	3	½	3½

	Experience in Years		
Grade	General	Specialized	Total
GS-7	3	1	4
GS-8	3	1½	4½
GS-9	3	2	5
GS-11&12	3	3	6

In addition, the applicants must show an ability to work without supervision and to write reports. They must be able to pass a physical examination and be free from color blindness. A technician might, in connection with their performance of regular duties, be required to drive a government-owned vehicle.

Opportunities for Training. Basic training is available at technical and vocational schools offering courses in electronics. Upon assignment to an FAA sector office, the new employees undergo a short period of on-the-job training to familiarize them with FAA equipment and procedures and then might receive several months of training at the FAA Academy in Oklahoma City.

The Academy offers correspondence courses to support technical training efforts, and many of these correspondence courses are prerequisites to assignment for advanced courses at the Academy. The technician receives regular salary and a subsistence allowance while in training at the Academy. Basic training and experience for FAA employment can also be obtained during active duty in the military service.

Electronics Technician—Airspace System Inspection

Technicians appointed for airborne technical/electronics duty are required to fly in government aircraft as a member of a crew with airspace system inspection pilots for data collection, evaluation, and/or engineering purposes during the in-flight inspection of navigational aids. Applicants should indicate on their applications their willingness to fly. Basically, the requirements and salary scales are the same as for the electronic technician.

Aviation Safety Inspector (General Information)

Aviation safety inspectors develop, administer, and enforce regulations and standards concerning civil aviation safety including (1) the airworthiness of aircraft and aircraft systems; (2) the competence of pilots, mechanics, and other personnel, and (3) safety aspects of aviation facilities, equipment, and procedures. These positions require knowledge and skill in the operation, maintenance, or manufacture of aircraft and aircraft systems. Aviation safety inspectors are categorized into three areas: *operational, airworthiness*, and *manufacturing*.

Physical Requirements. Candidates must be physically able to perform efficiently the duties of the position. They must have good distant vision in each eye and be able to read without strain printed material the size of typewritten characters; glasses are permitted. Ability to hear the conversational voice with or without a hearing aid is required.

Any physical condition would cause the applicant to be a hazard to himself or herself or others or that would interfere with her or his ability to fly as a passenger in a variety of airplanes will disqualify the applicant for appointment. In addition, candidates for positions must possess a currently valid first-class medical certificate in accordance with the regulations of the FAA. Incumbents of these positions must pass recurrent medical examinations as prescribed by the FAA.

Basis of Rating. No written examination is required. Candidates are rated, on a scale of 100, on the extent and quality of their experience and training. Ratings are based on candidate's statements in their applications and additional information that might be obtained by the Office of Personnel Management. For positions that involve specialization in flight operations, the nature, amount, and recency of flight time as a pilot or flight instructor is given substantial weight in ranking candidates.

Interview. Before appointment, candidates might be required to appear for an interview. The purpose is to observe and evaluate certain personal characteristics to determine whether candidates possess the following essential qualities:

- Ability to express ideas logically and accurately and to speak effectively and convincingly.
- Ability to operate successfully and easily in group situations. The following are elements of aviation safety positions that the candidates' background should have provided:
  - Broad knowledge of specialization
  - Independence and responsibility
  - Skill in evaluation and fact-finding
  - Ability to advise and guide others
  - Skill in reading comprehension and report writing

Selective Placement. For positions that require particular knowledge and skills, consideration might be restricted to those candidates whose background indicates that they possess the necessary knowledge and skills. For example, the Aviation Safety Inspector in an operational position might require ability to operate a specific type of jet aircraft or helicopter, in which case consideration might be restricted to candidates who have ratings in that type of aircraft. As another example, positions that primarily require knowledge and skill in maintenance of avionics equipment may be restricted to candidates whose backgrounds demonstrate knowledge and skill in the avionics area. Alternatively, separate registers could be established for eligibles with avionics expertise.

Working Conditions. The jobs require considerable travel, because inspections, consultations, and investigations must be made at various facilities and locations or at the scenes of accidents. Forty hours constitute a normal work week. Change of assignment from one duty station to another is required as staffing demands.

Where the Jobs Are. Inspectors operate out of Air Carrier District Offices, General Aviation District Offices (GADOs) and Flight Standards District Offices (FSDOs). These are located throughout the country. Five International Field Offices have the same functions as the FSDOs.

Opportunities for Advancement. Outstanding inspectors might be promoted to the next higher level with increased responsibilities and salary. An inspector demonstrating managerial ability could become a section or branch chief. She or he could also become an instructor at the FAA Academy.

Outlook for the Future. The recent recession and associated government hiring freeze has resulted in a net loss of inspectors, but concern over safety since the advent of economic deregulation of the airlines has put new emphasis on the need for more aviation safety inspectors.

AVIATION SAFETY INSPECTOR—OPERATIONS

Nature of the Work. Persons appointed to these positions apply knowledge and skills acquired as airmen (pilots, navigators, flight instructors, etc.) to develop and administer the regulations and safety standards pertaining to the operation of aircraft. Their primary duties include (1) examining airmen for initial certification and continuing competence; (2) evaluating airmen training programs, equipment, and facilities; and (3) evaluating the operation aspect of programs of air carriers and other commercial aviation operations.

Requirements to Enter the Job. Examples of qualifying specialized experience are:

- Pilot (or copilot) experience that provided comprehensive knowledge of operations requirements, facilities, practices, procedures, and flight activities of scheduled or supplemental air carriers, commercial operators, executive operators, air taxis, air travel clubs, or other civilian or military activities.
- Flight instructor in civilian or military training school.
- Flight test pilot or flight inspector involved in enforcing regulations concerning the safe operation of aircraft.
- Aviation operations inspector.

Certificates and Ratings.

- *Operations—General Aviation:* Candidates for positions that require operation of aircraft in the general aviation field must hold a current Commercial Pilot Certificate and current Flight Instructor Certificate, each with single-engine and multiengine land and instrument ratings.
- *Operations—Air Carrier.* Candidates for positions that require flight operation of aircraft in the air carrier field must (1) hold a current Airline Transport Pilot Certificate, or (2) hold a current Commercial Pilot Certificate with multiengine land and instrument ratings and be eligible for an Airline Transport Pilot Certificate.

Opportunities for Training. Flight training can be obtained in the military service or at privately or university-operated flight schools for commercial pilot's license and for multiengine and instrument ratings. From time to time, retraining is required as new developments in aircraft engines and equipment appear.

Aviation Safety Inspector—Airworthiness

Nature of the Work. Persons appointed to these positions apply knowledge and skills acquired as repair persons of aircraft and aircraft parts or avionics equipment to develop and administer regulations and safety standards pertaining to the airworthiness and maintenance of aircraft and related systems. Their primary duties include:

- evaluating mechanics and repair facilities for initial certification and continuing adequacy
- evaluating mechanics training programs
- inspecting aircraft and related systems for airworthiness
- evaluating the maintenance aspects of programs of air carriers and other commercial operators including the adequacy of maintenance facilities, equipment and procedures, the competence of personnel, the adequacy of the program or schedule for periodic maintenance and overhauls, and the airworthiness of aircraft.

Requirements to Enter the Job. Examples of qualifying experience are:

- Experience involving technical supervision or management of the maintenance and repair of aircraft, aircraft engines, or aircraft electronics communication and navigation systems and equipment of aircraft with responsibility for airworthiness following federal aviation or military regulations and safety standards. This experience must demonstrate a broad and comprehensive knowledge of maintenance of aircraft or aircraft systems. It must also demonstrate an ability to gain cooperation at management levels in complying with and supporting proper maintenance standards.
- Aviation safety inspector or air safety investigator concerned with aircraft powerplants, structures or systems.
- Experience gained as a field service representative of a manufacturer of aircraft systems and equipment may be accepted up to a maximum of one year.

Certificates and Ratings. Candidates for position at GS-9 and above are required to have knowledge and skill in the maintenance of aircraft (except positions that involve primarily avionics equipment) and must hold an FAA Mechanic Certificate with airframe and powerplant ratings.

Opportunities for Training. Basic training as an aircraft mechanic or in electronics and communications systems repair in a technical or vocational school is a starting point. College level work in aeronautical engineering or aeronautical maintenance or electrical or electronic engineering is preferred. From time to time, retraining is required as new developments in aircraft, engines, and equipment appear.

Aviation Safety Inspector—Manufacturing

Nature of the Work. Persons appointed to these positions apply knowledge and skills pertaining to the design and production of aircraft, aircraft parts, and avionics equipment to develop and administer regulations and safety standards pertaining to the original airworthiness of aircraft, aircraft parts, and avionics equipment. Their primary duties include:

- ✦ Inspecting prototype or modified aircraft, aircraft parts, and avionics equipment for conformity with design specifications
- ✦ Inspecting production operations including equipment, facilities, techniques, and quality control programs for capability to produce the aircraft or parts in conformance with design specifications and safety standards
- ✦ Making original airworthiness determinations and issuing certificates for all civil aircraft including modified, import, export, military surplus, and amateur-built aircraft.

Requirements to Enter the Job. Examples of qualifying specialized experience are:

- ✈ Experience involving quality control of the manufacture of aircraft, aircraft engines, propellers, or major aircraft assemblies produced under the requirements of federal aviation regulations.
- ✈ For grades GS-11 and above, this experience must have demonstrated the ability to evaluate and provide technical guidance and direction to the quality control program of a manufacturer producing aircraft, aircraft engines, propellers, or major aircraft assemblies. This experience may have been acquired in such positions as quality control engineer, quality control supervisor, or service representative with quality control supervisory experience.

Opportunities for Training. A college degree in aeronautical, production, or industrial engineering is the best preparation for entry into jobs at higher levels. Technical or vocational school training in various trades associated with aircraft manufacturing (drafting, sheet metal work, air conditioning, electrical systems, etc.) leading to jobs in aircraft manufacturing can give minimum background and experience. From time to time, retraining is required as new developments in aircraft, engines, and equipment appear.

AIRSPACE SYSTEM INSPECTION PILOT

Nature of the Work. The airspace system inspection pilots conduct in-flight inspections of ground-based air navigational facilities to determine if they are operating correctly. They pilot multiengine, high-performance jet aircraft with specially installed, ultra-sophisticated, computerized, and automated electronic equipment to serve as a flying electronic laboratory. They fly on day and night flights, under both visual and instrument flight rules, recording and analyzing facility performance and reporting potential hazards to air navigation for correction. The pilot assists in accident investigations by making special flight tests of any FAA navigational aids involved. He or she maintains liaison with aviation interests outside the FAA regarding the installation, operation, and use of their navigation facilities, but these pilots are mostly involved with the FAA employees who maintain the navaids.

Working Conditions. The job requires considerable travel, as flights cover navigation aids supporting federal airways and civil and military airports located throughout the United States. The basic work week consists of forty hours.

Where the Jobs Are. Inspection pilots work out of one of seven Flight Inspection Field Offices within the coterminous 48 states. Upon reaching the journeyman level of proficiency, you could, at your option, bid on a job in one of the flight inspection offices in Alaska, Hawaii, Tokyo, or Germany.

Wages. The entry level is GS-9.

Opportunities for Training. An employee enters as a trainee and then advances to the job of Second Pilot on an in-flight inspection or at air navigation facilities. The next step is that of Supervisory Airplane Pilot who supervises the flight inspection crew and evaluates the report findings on navigation systems. The top jobs, located in field offices, are those of supervisors responsible for the overall program accomplishment of the field offices. If assigned to a Flight Inspection Field Office, the employee can advance through Second Pilot and to Airspace and Procedures Specialist and be responsible for developing instrument approach, terminal, and enroute air traffic procedures. Or, he or she could move up to become Senior Flight Inspector and Aircraft Commander, supervising flight crews and being responsible for results of inspection missions. Managers of the field offices are the top jobs.

Requirements to Enter the Job. Experience as a pilot in general is required. Experience requirements are specified in terms of flying time and certificates and ratings rather than in the number of years of experience. As a minimum, she or he must hold a valid commercial pilot certificate with multiengine rating and instrument ratings. See the following table for requirements.

	Total Time	Pilot-In-Command	Multi-Engine	Instrument Night	Last 12 Months
GS-9	1200	250	100	100	100
GS-11	1500	250	500	150	100

The flying time in any category (except the pilot-in-command column) may be as pilot or copilot. The instrument/night requirement must include at least 40 hours of actual instrument weather time. Experience as an air traffic controller, chief test pilot, chief pilot of an FAA certificated flight school, or designated pilot examiner may be substituted for not more than 50 hours of the flying time required for the last 12 months. The pilot must have a valid first-class FAA medical certificate and must requalify periodically in physical examinations to maintain employment in this job.

Opportunities for Training. Flight instruction may be obtained from privately or university-operated flight schools or from the military services.

FLIGHT TEST PILOT

Nature of the Work. The FAA flight test pilot checks the airworthiness of aircraft through inspection, flight testing, and evaluations of flight performance, engine operation, and flight characteristics of either prototype aircraft or modifications of production aircraft and aircraft components that are presented for FAA-type certification. The flight test pilot supervises FAA-designated flight-test representatives and participates in investigations of accidents and violations of the Federal Air Regulations.

Working Conditions. This employee flies new types of aircraft under all kinds of conditions to test their performance. Considerable travel is required and his or her duty station might be changed from time to time as circumstances require.

Where the Jobs Are. The jobs are located in areas where aircraft manufacturing plants are situated. They are chiefly in California, Washington, Missouri, Maryland, Texas, Kansas, Florida, and New York.

Wages. The entry level is GS-9. Entrance salary varies with the applicant's experience and training.

Opportunities for Advancement. The flight test pilot can progress to branch chief positions in the engineering or manufacturing areas. An administrative post with respect to all FAA flight test pilots at FAA Headquarters or perhaps an assignment with the FAA Test Center (the research and development arm of FAA) might provide opportunities for an administrative flight test engineering job.

Requirements for the Job. Three years of general experience as a pilot or copilot in any civilian or military major aircraft operation is required. Also required is one to three years of special experience in the aircraft manufacturing industry or in the military or civil service of the federal government as a flight test pilot, aeronautical engineer, or flight test engineer. The special experience must include engineering flight testing of experimental types of aircraft or the solution of technical engineering problems at a professional level. The pilot must have experience in obtaining and evaluating flight data related to flight performance, flight characteristics, engine operation, and other performance details of the prototype or modifications of production aircraft. Experience as an instructor in engineering flight testing of aircraft is also required. The higher entry grades require completion of a flight test pilot course, such as a military flight test school or the FAA flight test pilot course. College study in aeronautical, electrical, electronic or mechanical engineering, mathematics or physics may be substituted for some of the general experience requirements. He or she must have a first-class FAA medical certificate plus 1,500 to 2,000 hours of flight time, a commercial pilot license, and single engine, multiengine, and instrument ratings. She or he must pass physical exams at regular intervals to retain the job.

Opportunities for Training. Obtain flight training with advanced training at a military flight test school in the military service. Or, you can get the initial flight training through commercial pilot's license with appropriate ratings from privately or school-connected flying schools and institutes. A college degree in aeronautical engineering with flight training is preferred.

MAINTENANCE MECHANIC

Nature of the Work. There are a number of employees classified under the federal wage system. These employees perform jobs associated with the trades and crafts and are paid on an hourly basis. One example is the FAA maintenance mechanic. They maintain aids to navigation such as the approach light systems serving airport runways. They also work on the structural, electrical, and mechanical devices that are major parts of other facilities. This includes maintenance and repair of heating, air conditioning, and ventilating systems; electrical generating and power distribution systems; and the buildings and antenna structures that house a wide variety of FAA facilities. The job involves carpentry, painting, plumbing, electrical, and masonry construction, and installation, repair, and maintenance of air conditioning, heating, or power-generating equipment.

Working Conditions. Work is indoors or outdoors as the jobs require. Work could be on outdoor structures of heights up to 300 feet. The basic work week consists of 40 hours. The employee must be able to drive a truck to jobs in outlying areas.

Where the Jobs Are. Such jobs are located in all areas of the U.S., Puerto Rico, the Virgin Islands, and anywhere that FAA air navigational aids and air traffic control towers and centers are situated.

Wages. Hourly wages vary according to experience and the prevailing rates paid where the jobs are located.

Requirements to Enter the Job. The employee must have four years of progressively responsible experience in two or more of the following occupational groups: machinist, machine repair person, automobile mechanic, carpenter, woodworker, electrician, electric motor repair person, painter, air conditioning and refrigeration repair person, heating equipment, and power-generating equipment repair person. Training in a trade school may be substituted for some of the required experience. Candidate must have a driver's license.

Opportunities for Training. Training can be acquired in high school industrial arts classes and in vocational or technical schools.

ENGINEER

General Information. The FAA, as well as the National Aeronautics and Space Administration and the Department of Defense employs engineers of all specialties to work on research and development problems in aviation, such as V/STOL (very short takeoff and landing) aircraft, aircraft sound, the sonic boom, hypersonic aircraft and new equipment and devices to increase aviation safety. Engineers also provide guidance in airport design, construction, operations and maintenance.

Nature of the Work. The facilities, devices, and machines needed by the FAA to carry on its work require the services of a number of engineering specialists.

- *Aerospace (Aeronautical) Engineers* develop, interpret, and administer safety regulations relating to airworthiness of aircraft and their accessories. They analyze and evaluate manufacturers' designs, set up test procedures, observe tests, and furnish engineering advice to manufacturers. They deal with such problems as vibration, flutter, stability, control, weight and balance, aerodynamic characteristics, etc.
- *Electrical Engineers* deal with power supply, distribution, and standby power generation required for the operation of air navigational aids. They are also involved in the design and evaluation of airport and runway lighting and electrical equipment aboard aircraft.
- *Electronic Engineers* are concerned with designing improved electronic navigational aids and communications systems. They may design, develop, modify, or oversee installation, calibration, and maintenance of ground and airborne electronic equipment. They recommend location of aids.
- The Mechanical Engineers are concerned with the design of gasoline and diesel power plants for standby power generation in case of emergencies. They are also concerned with heating, ventilating, and air conditioning equipment at FAA installations. Some mechanical engineers check out such things as the performance of new types of aircraft engines, fuel systems, and fire detection devices.
- The Civil Engineer involved in the airports program deals with a broad range of airport design, construction, and maintenance matters. FAA involvement in these matters is in the area of providing advice and guidance to civil airport developers with particular emphasis on airports developed with federal grants in aid.

Working Conditions. The engineer works at a desk in an engineering laboratory or outdoors conducting or observing tests of equipment during a forty-hour week. Travel might be required as the engineer consults with aircraft and engine manufacturers and with suppliers of all kinds of equipment related to the engineering specialty. Engineers may travel to consult with state and city officials who need Federal funds for building or improving airports and to military bases where equipment is tested.

Where the Jobs Are. Engineering jobs are located at FAA headquarters, district, and regional offices, at NASA Headquarters and centers, and at certain military bases scattered throughout the nation.

Wages. GS-5 to GS-14 are beginning salaries, depending upon previous experience and education background.

Opportunities for Advancement. Promotion is normally from within.

Requirements of the Job. A B.S. degree in engineering or four years of technical engineering experience and training is required that provides technical knowledge equal to that possessed by a graduate engineer. Zero to three additional years of experience are required, depending on entry grade level.

Opportunities for Training. Engineering training is available from colleges offering courses in the various specialized engineering fields.

ENGINEERING AID OR TECHNICIAN

Nature of the Work. Depending on the specialty, the engineering aid or technician assists engineers by drafting engineering plans, conducting efficiency and performance tests, making calculations, setting up laboratory equipment and instruments, and preparing technical reports, specifications, and estimates.

Working Conditions. The basic work week is 40 hours. Travel might be required as the technician consults with aircraft and engine manufacturers and with suppliers of all kinds of equipment related to her or his engineering specialty. He or she might have to travel to consult with state and city officials who need Federal funds for building or improving airports and to military bases where equipment is tested.

Where the Jobs Are. The jobs are located at FAA facilities and at FAA's Test Center at Atlantic City, NJ, at NASA Headquarters and centers, and at certain military bases scattered throughout the nation.

Wages. The starting salaries for engineering aids are GS-1 to GS-3 and for engineering technicians GS-4 to GS-12, depending upon previous experience and educational background.

Requirements to Enter the Job. See the following table for experience requirements.

	Experience in Years		
Grade	General	Specialized	Total
GS-1	0	0	0
GS-2	½	0	½
GS-3	1	0	1
GS-4	1½	½	2
GS-5	2	1	3
GS-6	2	2	4
GS-7	2	3	5
GS-8 and above	2	4	6

Engineering technicians can be certified by the Institute for Certification of Engineering Technicians, an organization sponsored by the National Society of Professional Engineers. They can be certified as Junior Engineering Technician, Engineering Technician, or Senior Engineering Technician.

Opportunities for Training. Attend a vocational or technical school, junior/community college, or a four-year college.

OTHER PROFESSIONAL EMPLOYEES OF THE FAA

The FAA also requires the services of professional people other than engineers. Aviation medicine is a most important function, and physicians who have chosen aviation medicine as a specialty beyond their general medical education are employed by the FAA in limited numbers. These physicians study such things as the effects of flying on the human body, the need for oxygen above certain altitudes, the effects of fatigue on pilot performance, vision and hearing standards, the tension and stress factors associated with the air traffic controller's job, and the standards of the various classes of medical examinations required for pilots and other members of flight crews.

A wide scope of professionals are represented in FAA. These include airport safety specialists, urban planners, economists, mathematicians, statisticians, program officers, management analysts, and budget analysts.

The FAA requires logistic support for all its programs, particularly in the area of establishment, operation, and maintenance of air navigation and traffic control facilities. Personnel plan and manage programs for the establishment and installation of facilities, acquire real and personal property, operate and maintain facilities, and provide all aspects of property and material management. To do so, FAA employs personnel such as logistics program planners and managers, real property specialists, inventory and supply managers, procurement analysts, contracting specialists, transportation officers, and purchasing clerks.

The FAA also employs lawyers to write Federal Aviation Regulations, to interpret them, and to represent the FAA in legal controversies. It employs public information officers, librarians, photographers, and supporting personnel such as receptionists, secretaries, typists, office machine operators, mail room clerks, and computer programmers and operators. In addition, the FAA operates two Federal Commercial airports in the Washington, D.C. area and for this facility employs runway, building, and ground maintenance personnel as well as an airport administrative staff.

NATIONAL TRANSPORTATION SAFETY BOARD

The National Transportation Safety Board accident investigators interview survivors and witnesses and examine aircraft parts, instruments, and engines. They also review maintenance and flight records to determine the probable cause of airplane accidents. Aviation-related engineering, medical, and/or operational experience is required for a variety of professional positions with this safety-related organization. Travel and field work typify the investigator's position. Salary and experience rankings resemble those of the Department of Transportation.

U.S. MILITARY SERVICES

The U.S. military services are large employers of civilians for jobs in aviation such as aircraft mechanics, engineers, technicians, and general office workers such as secretaries and typists. These civilian jobs come under the Federal Civil Service, and employees do many of the same kinds of work and receive the same wages and benefits as their counterparts in the FAA or other federal government departments and agencies.

There are many aviation career opportunities in the military services for enlisted personnel and officers, both men and women. The Air Force, of course, offers the greatest number of training and employment opportunities to fly as a pilot or to work as an aircraft mechanic, air traffic controller, electronic technician, flight nurse, or meteorological technician, to name a few. Navy and Marine aviation also have their counterpart jobs to those in the Air Force. Army aviation is mostly connected with operation and maintenance of helicopters and subsonic light planes and requires flight crews, ground service people, and weather specialists to support its aerial operations. Many of these military aviation jobs prepare the service man or woman for similar jobs in civilian life if he or she chooses to leave the service. For example, flying officers released by the military services especially the Air Force, have constituted the major source of supply of airline pilots in recent years. An Air Transport Association survey revealed that a high percentage of all airline pilots employed had their principal training in the military service.

OTHER FEDERAL GOVERNMENT DEPARTMENTS AND AGENCIES

Many other federal government departments, bureaus, and agencies operate aircraft to carry on their work more effectively. For example, the Fish and Wildlife Service of the Department of the Interior uses airplanes to make wildlife census. The Department of Agriculture's Forest Service uses aircraft to check on aerial forest spraying contracted to commercial operators or to oversee forest firefighting procedures. The Immigration and Naturalization Service of the Department of Justice uses aircraft to detect people entering the U.S. illegally, and the U.S. Coast Guard operates aircraft for search and rescue purposes. Although pilot and mechanic jobs within these agencies are comparatively few in number, they are mentioned to complete the full picture of aviation career opportunities within the Federal Civil Service.

Pilots for these federal government agencies fly aircraft to transport office staff members and supplies, perform aerial surveys, make wildlife census, etc. as required by their particular government office. They fly in single or multi-engine aircraft during day or night, as required, and over all kinds of terrain in all kinds of flyable weather.

The jobs are based throughout the country wherever the department operations require. Pilots must have from 1,200 to 2,500 flying hours experience including extended cross-country flights over land and/or water during which they perform their own navigating. They must be able to pass a first-class or a second-class FAA physical examination every 6 or 12 months, respectively. The annual salary ranges from GS-9 to GS-12, depending on experience and educational background.

NATIONAL WEATHER SERVICE METEOROLOGIST

Aviation is one of the largest consumers of weather information. Flight and weather are so interrelated that many people in aviation look upon the weather person as a member of the aviation team. Thus, the meteorologist deserves mention in any discussion of vocations in aviation, even though these functions are not, of course, entirely for the benefit of the aviation community.

Nature of the Work. In general, the meteorologist who works most closely with aviation is an operational, or *synoptic*, meteorologist (as contrasted to the meteorologist working in theoretical or applied meteorological research). He or she is the forecaster who provides the day-to-day, hour-to-hour observations, analyses, forecasts, warnings, and advice to pilots, airport operators, and airlines. She or he reports weather conditions expected at airports, current conditions, and enroute forecasts.

The meteorologist's main tasks involve the interpretation of meteorological data provided by weather observations and instruments. At a small weather station, he or she might carry on numerous functions such as: making outside weather observations, reading and recording data from weather instruments, checking weather data coming in via a machine, drawing weather maps, plotting the weather, providing forecasts, and advising pilots and other interested parties. At larger stations, the meteorologist might specialize in one or more of these duties, relying to some extent upon computerized data in order to produce a forecast. She or he sends forecasts via teletype or telephone to Flight Service Stations, airline dispatch offices, airports, and to other consumers of weather information. Often, the meteorologist advises pilots personally when assisting the pilot in drawing up a flight plan.

Working Conditions. The meteorologist works indoors, sitting or standing at map tables while working on weather maps and charts. He or she reads data from weather instruments such as anemometers, thermometers, barometers, theodolites, ceilometers, radiosondes, weather balloons, etc. The meteorologist is able to operate a teletypewriter. At times, she or he might be required to work outdoors for short periods, checking weather instruments and making observations. He or she might work alone at a small station or with other meteorologists and meteorological technicians at a large station. At airport Weather Service Stations, he or she meets with private, business, and airline pilots. Meteorologists usually work a 40-hour week. Overtime is required when weather conditions deteriorate. Shift work is required when a station is open 24 hours a day. A meteorologist might be required to relocate to fill staffing requirements at another station or to advance in grade.

Where the Jobs Are. The largest employer of federal government meteorologists is the National Weather Service. Several thousand Weather Service meteorologists work at approximately 300 stations scattered throughout the 50 states, in Puerto Rico, in Arctic regions, at Wake Island, and at other Pacific Ocean sites. Major Weather Service Stations are located at airports or in large cities. A smaller number of federal government meteorologists work for the Air Force, Navy, Army, the FAA, NASA, and the U.S. Forest Service.

The Weather Service also employs meteorological technicians to assist meteorologists. Most of the job vacancies for this position are filled by applicants who have received their technical training during active duty in the Armed Forces. The meteorological technician performs semiprofessional and scientific work, calibrating and using instruments for taking various kinds of measurements, observing, recording, computing, processing, classifying, and disseminating weather data.

Wages. The salary ranges from an annual starting grade of GS-5 to a GS-15. The starting salary and grade is determined by the amount of education and experience.

Opportunities for Advancement. Promotion to higher grades depends upon education, ability, work performance. In-grade pay increases are made on the basis of experience and satisfactory performance of the job. With an increase in grade comes increased responsibilities as assistant chief or chief of a weather station or region. A few high administrative jobs are available as vacancies occur.

Requirements to Enter the Job. One of the following two items is required:

- A full course of study leading to a bachelor's degree at an accredited college or university that has included or been supplemented by 20 semester hours in meteorology (including 6 semester hours in weather analysis and forecasting and 6 semester hours in dynamic meteorology), and in addition, differential, and integral calculus and 6 semester hours in college physics.
- At least 20 semester hours in meteorology at an accredited college or university that has 6 semester hours in weather analysis and forecasting and 6 semester hours in dynamic meteorology. Also, differential and integral calculus and 6 semester hours in college physics plus additional appropriate education or technical experience that when combined with the education prescribed above will total four years of education or education and experience. This pre-professional background must be of such quality that it provides a body of knowledge and abilities comparable to that normally acquired through the successful completion of a full course of study described in paragraph one above.

Opportunities for Training. More than 20 universities offer bachelor degrees in meteorology or equivalent, with others offering a major in meteorology. Training as a meteorological technician is obtainable while on active duty with the armed services or at some junior/community colleges or technical institutions. The Weather Service operates a Technical Training Center in Kansas City for the purpose of upgrading technicians.

Outlook for the Future. The science of meteorology is expanding and with it comes increased occupational opportunities. The Weather Service expects to hire at least 100 meteorologists with a B.A. degree each year to fill new positions and vacancies. Opportunities for military careers in meteorology are excellent and competent military meteorologist officers are given opportunities to receive advanced degrees at government expense. At present, the number of qualified students obtaining degrees in meteorology are fewer than can supply future expected demands. Although the demand is small, so few are entering the occupation that job opportunities are available for the qualified applicant.

STATE AVIATION JOBS

Almost every state has an Aeronautics Department or Commission that consists of a small number of aviation-minded men and women usually appointed by the Governor to make policies about aviation activities within the state. Sometimes, persons appointed are not considered state employees and might only receive payment for expenses connected with their attendance at meetings. If the State Department or commission is well funded, it has some employees who work in the areas of airport design and operation, flight safety, and promotion of aviation activities in the state.

Frequently, employees have dual responsibilities, especially when the staff is small. Qualifications and requirements for these various jobs are determined by state law; however, the top-level employees (safety officers, field service representatives, and engineers) must have experience and training in their specialty. Almost all employees working under state civil service enjoy retirement plans, social security, low-cost insurance, and medical service. In most cases, State Department employees work out of the office in the State Capitol. Typical state aviation positions are shown in Table 5-2.

OUTLOOK FOR AVIATION CAREER OPPORTUNITIES IN THE GOVERNMENT

The outlook for growth in aviation career opportunities with the federal government is mixed. Rapid increases in automation have lessened the need for increased numbers of air traffic controllers. This fact, combined with the recent recession and concern about budget deficits, will no doubt keep employment levels static. It should be pointed out, though, that normal attrition from retirements, etc. will allow hiring to continue at its present pace for the foreseeable future. Current emphasis on decreasing federal participation in the economy will mean that with economic recovery, federal aviation employment will not expand to the degree that civil aviation employment will.

The future of civilian aviation careers with the military services is somewhat uncertain as demand is responsive to world conditions. The best predictions indicate at least a small increase over the next ten years.

Employment in aviation at the state government level will likely show an upward trend as aviation activities within the state grow in proportion to decreases in federal activity.

For information about job openings at the federal government level, call the Federal Job Information Service (FJIS) closest to the area of desired employment. In most cases, you will reach a prerecorded message that lists the positions that are vacant and being staffed through that office. Table 5-3 shows a listing of the addresses and telephone numbers of all of the Offices of Personnel Management (OPMs) for federal job information, categorized by state.

For information about job openings at the state level, write or call the appropriate State Aeronautical Agency. The personnel division in the agency is responsible for staffing vacant positions within the organization. See Table 5-4 for the addresses and telephone numbers of State Aeronautical Agencies.

For information concerning job openings with the FAA, write to the Federal Aviation Administration Regional Office that covers the desired area of employment. There are ten Regional Offices in the U.S. as broken down in Table 5-5.

Table 5-3. FEDERAL OFFICE OF PERSONNEL MANAGEMENT JOB INFORMATION/TESTING OFFICES

ALABAMA

Huntsville:
Southerland Building
806 Governors Dr., S.W., 35801
(205)544-5802

ALASKA

Anchorage:
Federal Building
701 C.St., Box 22, 99513
(907)271-5821

ARIZONA

Phoenix:
U.S.Postal Service Building
522 N. Central Ave.,
Room 120, 85004
(602)261-4736

ARKANSAS

Little Rock:
Federal Building, Rm. 3421
700 W. Capitol Ave., 72201
(501)378-5842

CALIFORNIA

Los Angeles:
Linder Building, 3rd Floor
845 S. Figueroa, 90017
(213)894-3360

Sacramento:
1029 J St., 2nd Floor, 95814
(916)551-1464

San Diego:
880 Front St., Rm. 92188
Federal Building, Rm. 459
(619)575-6165

San Francisco:
211 Main St.
Second Floor, Room 235, 94105
(415)974-9725

COLORADO

Denver:
P.O. Box 25167, 80225
(303)236-4160
Physically located at
12345 W. Alameda Pkwy.,
Lakewood, CO

For job information (24 hrs. a day)
in the following states, dial:
MONTANA: (303)236-4162
UTAH: (303)236-4165
WYOMING: (303)236-4166

For forms and local supplements, dial:
(303)236-4159

CONNECTICUT

Hartford:
Federal Building, Rm. 613
450 Main St., 06103
(203)240-3263

DELAWARE
(See Philadelphia, PA listing)

DISTRICT OF COLUMBIA

Metro Area:
1900 E St., N.W.
Rm. 1416, 20415
(202)653-8468

FLORIDA

Orlando:
Commodore Building, Suite 125
3444 McCrory Pl., 32803-3712
(305)648-6148

GEORGIA

Atlanta:
Richard B. Russell Federal Building
Rm. 960, 75 Spring St., S.W., 30303
(404)331-4315

GUAM

Agana:
Pacific Daily News Building
238 O'Hara St., Rm. 902, 96910
472-7451

HAWAII

Honolulu (and other Hawaiian Islands and overseas):
Federal Building, Rm. 5316
300 Ala Moana Blvd., 96850
(808)541-2791
(808)541-2784 (Overseas Jobs)

IDAHO
(See Washington listing)

ILLINOIS

Chicago:
175 W. Jackson Blvd.
Room 530, 60604
(312)353-6192

INDIANA

Indianapolis:
Minton-Capehart Federal Building
575 N. Pennsylvania Ave., 46204
(317)269-7161

IOWA
(See Kansas City listing)

KANSAS

Wichita:
One-Twenty Building, Rm. 101
120 S. Market St., 67202
(316)269-6796

In Johnson, Leavenworth and Wyandotte counties, dial
(816)374-5702

KENTUCKY
(See Ohio listing)

LOUISIANA

New Orleans:
F. Edward Hebert Building
610 S. Maestri Pl., Rm. 802, 70130
(504)589-2764

MAINE
(See New Hampshire listing)

MARYLAND

Baltimore:
Garmatz Federal Building
101 W. Lombard St., 21201
(301)962-3822

MASSACHUSETTS

Boston:
Boston Federal Office Building,
10 Causeway St., 02222-1031
(617)565-5900

MICHIGAN

Detroit:
477 Michigan Ave.,
Rm. 565, 48226
(313)226-6950

MINNESOTA

Twin Cities:
Federal Building
Ft. Snelling, 55111
(612)725-3430

MISSISSIPPI
(See Alabama listing)

MISSOURI

Kansas City:
Federal Building, Rm. 134
601 E. 12th St., 64106
(816)374-5702

St. Louis:
Old Post Office, Rm. 400
815 Olive St., 63101
(314)425-4285

MONTANA
(See Colorado listing)

NEBRASKA
(See Kansas listing)

NEVADA
(See Sacramento, CA listing)

NEW HAMPSHIRE

Portsmouth:
Thomas J. McIntyre Federal Building, Room 104
800 Daniel St., 03801-3879
(603)431-7115

NEW JERSEY

Newark:
Peter W. Rodino, Jr., Federal Building,
970 Broad St., 07102
(201)645-3673

In Camden, dial (215)597-7440

NEW MEXICO

Albuquerque:
Federal Building
421 Gold Ave., S.W., 87102
(505)766-5583

In Dona Ana, Otero, and El Paso counties, dial (505)766-1893

NEW YORK

New York City:
Jacob K. Javits Federal Building
26 Federal Plaza, 10278
(212)264-0422

Syracuse:
James N. Hanley Federal Building
100 S. Clinton St., 13260
(315)423-5660

NORTH CAROLINA

Raleigh:
Federal Building
310 New Bern Ave.
P.O. Box 25069, 27611
(mailing address only)
(919)856-4361

NORTH DAKOTA
(See Minnesota listing)

OHIO

Dayton:
Federal Building, Room 506
200 W. 2nd St., 45402
(513)225-2720

OKLAHOMA

Oklahoma City:
(mail or phone only)
200 N.W. Fifth St.
Second Floor, 73102
(405)231-4948

OREGON

Portland:
Federal Building, Rm. 376
1220 S. W. Third St., 97204
(503)221-3141

PENNSYLVANIA

Harrisburg:
Federal Building, Rm. 168
P.O. Box 761, 17108
(717)782-4494

Philadelphia:
Wm. J. Green, Jr., Federal Building
600 Arch St., Rm. 1416, 19106
(215)597-7440

Pittsburgh:
Federal Building
1000 Liberty Ave., Rm. 119, 15222
(412)644-2755

PUERTO RICO

San Juan:
Federico Degetau Federal Building
Carlos E. Chardon St.
Hato Rey, P.R. 00918
(809)753-4209

RHODE ISLAND

Providence:
John O. Pastore Federal Building
Rm. 310, Kennedy Plaza, 02903
(401)528-5251

SOUTH CAROLINA
(See North Carolina listing)

SOUTH DAKOTA
(See Minnesota listing)

TENNESSEE

Memphis:
200 Jefferson Ave.
Suite 1312, 38103-2355
(901)521-3956

TEXAS

Dallas:
(mail or phone only)
Rm. 6812, 1100 Commerce St., 75242
(214)767-8035

Houston:
(phone-recording only)
(713)226-2375

San Antonio:
(mail or phone only)
643 E. Durango Blvd., 78206
(512)229-6611 or 6600

UTAH
(See Colorado listing)

VERMONT
(See New Hampshire listing)

VIRGINIA

Norfolk:
Federal Building, Rm. 220
200 Granby Mall, 23510-1886
(804)441-3355

WASHINGTON

Seattle:
Federal Building
915 Second Ave., 98174
(206)442-4365

WEST VIRGINIA
(See Ohio listing)
or dial (573)225-2866

WISCONSIN

Residents in Counties of Grant, Iowa, Lafayette, Dane, Green, Rock, Jefferson, Walworth, Waukesha, Racine, Kenosha and Milwaukee should dial (312)353-6189 for job information. All other Wisconsin residents should refer to the Minnesota listing for Federal job information in their area.

WYOMING
(See Colorado listing)

Table 5-4. STATE AERONAUTICAL AGENCIES

ALABAMA
Department of Aeronautics
Arthur Jones, Director
817 South Court St.
Montgomery, AL 36130
(205)261-4480

ALASKA
Department of Transportation and Public Facilities
Gina Lindsey, Manager
P.O. Box 6900
Anchorage, AK 99502

ARIZONA
Division of Aeronautics
Gary Adams, Acting Director
1801 West Jefferson, Room 426
Phoenix, AZ 85007
(602)255-7691

ARKANSAS
Department of Aeronautics
Tommy Hancock, Director
Little Rock Regional Airport
1 Airport Dr., 3rd Floor
Little Rock, AR 72202
(501)376-6781

CALIFORNIA
Division of Aeronautics
Jack Kemmerly, Chief
P.O. Box 942874
Sacramento, CA 94272-0001
(916)322-3090

COLORADO
Aviation Planning Staff
Philip Schmuck, Program Administrator
1313 Sherman St., Suite 520
Denver, CO 80203
(303)866-3004

CONNECTICUT
Bureau of Aeronautics
Robert H. Carrier, Chief Executive Officer
P.O. Drawer A
Wethersfield, CT 06109
(203)566-4417

DELAWARE
Aeronautics Administration
Rayvon Burleson, Administrator
P.O. Box 778
Dover, DE 19903
(302)736-3264

FLORIDA
Bureau of Aviation
Jack K. Johnson, Chief
605 Suwannee St. M.S. 46
Tallahassee, FL 32301-8064
(904)488-8444

GEORGIA
Bureau of Aeronautics
W. "Luke" Cousins III, Chief
2017 Flightway Dr.
Chamblee, GA 30341
(404)986-1350

HAWAII
Airports Division
Owen Miyamoto, Airports Administrator
Honolulu International Airport
Honolulu, HI 96819
(808)836-6432

IDAHO
Bureau of Aeronautics and Public Transportation
William Miller, Chief
3483 Rickenbacker St.
Boise, ID 83705
(208)334-3183

ILLINOIS
Division of Aeronautics
Robert Coverdale, Director
One Langhorne Bond Dr.
Springfield, IL 62707-8415
(217)785-8515

INDIANA
Division of Aeronautics
Wayne Reynolds, Deputy Director
143 W. Market St., Suite 300
Indianapolis, IN 46204
(317)232-1470

IOWA
Air and Transit Division
Nancy Richardson, Director
International Airport
Des Moines, IA 50321
(515)281-4280

KANSAS
Aviation Division, Department of Transportation
George M. Boyd, Director
State Office Building
Topeka, KS 66612-1586
(913)296-2553

KENTUCKY
Office of Aeronautics
Bob Bodner, Executive Director
US 127 South, Annex Building, Suite 1
Frankfort, KY 40622
(502)564-4480

LOUISIANA
Office of Aviation and Public Transportation
M.K. Johnston, DOTD Staff Engineer Coordinator
P.O. Box 94245
Baton Rouge, LA 70804-9245
(504)379-1235

MAINE
Division of Aeronautics
Linwood Wright, Acting Director
State Office Building
Augusta, ME 04333
(207)289-3185

MARYLAND
State Aviation Administration
Theodore Mathison, Administrator
P.O. Box 8766
Baltimore-Washington International Airport, MD 21240
(301)859-7060

MASSACHUSETTS
Aeronautics Commission
Arnold Stymest, Director
10 Park Plaza, Room 6620
Boston, MA 02116-3966
(617)973-7350

MICHIGAN
Aeronautics Commission
William E. Gehman, Director
2nd Floor, Terminal Building
Capital City Airport
Lansing, MI 48906
(517)373-1834

MINNESOTA
Office of Aeronautics
Raymond J. Rought, Director
Transportation Building
St. Paul, MN 55155
(612)296-8202

MISSISSIPPI
Aeronautics Commission
Kenneth A. Barfield, Director
P.O. Box 5
Jackson, MS 39205
(601)359-1270

MISSOURI
Department of Highways and Transportation
Lloyd D. Parr, Director of Aviation
P.O. Box 272
Jefferson City, MO 65101
(314)751-2589

MONTANA
Aeronautics Division
Michael D. Ferguson, Administrator
P.O. Box 5178
Helena, MT 59604
(406)444-2506

NEBRASKA
Department of Aeronautics
Anne G. Beaurivage, Director
P.O. Box 82088
Lincoln, NB 68510
(402)471-2371

NEVADA
Department of Transportation
Dennis O. Barry, Assistant Director
Planning and Programming
1263 South Stewart St.
Carson City, NV 89712
(702)885-5510

NEW HAMPSHIRE
Aeronautics Division
Harold W. Baker Jr., Director
65 Airport Rd.
Concord, NH 03301-5298
(603)271-2551

NEW JERSEY
Division of Aeronautics
Paul Baker, Acting Director
1035 Parkway Ave.
Trenton, NJ 08625
(609)530-2900

NEW MEXICO
Aviation Division
Merrill Goodwyn, Director
P.O. Box 579
Santa Fe, NM 87504-0579
(505)827-0332

NEW YORK
Aviation Division
Clarence M. Cook, Director
1220 Washington Ave.
Albany, NY 12232
(518)457-2820

NORTH CAROLINA
Division of Aviation
Willard G. Plentl, Jr., Director
P.O. Box 25201
Raleigh, NC 27611
(919)787-9618

NORTH DAKOTA
Aeronautics Commission
Gary R. Ness, Director
Box 5020, Bismarck Airport
Bismarck, ND 58502
(701)224-2748

OHIO
Bureau of Aviation
John B. Cornett, Administrator
2829 West Dublin-Granville Rd.
Worthington, OH 43235
(614)466-7120

OKLAHOMA
Aeronautics Commission
Jim C. Schooley, Director
200 N.W. 21st St., Room B-7, 1st Floor
Oklahoma City, OK 73105
(405)521-2377

OREGON
Division of Aeronautics
Paul E. Burket, Administrator
3040 25th St., S.E.
Salem, OR 97310
(503)378-4880

PENNSYLVANIA
Bureau of Aviation
Charles Hostetter, Director
Transportation and Safety Building
Room 716
Harrisburg, PA 17120
(717)783-2280

PUERTO RICO
Ports Authority
F. Guilleimo Valls, Esquire,
Executive Director
GPO Box 2829
San Juan, PR 00936
(809)723-2260

RHODE ISLAND
Division of Airports
E.A. Tansey, Assistant Director
of Transportation (Airports)
Theodore F. Green State Airport
2000 Post Rd.
Warwick, RI 02886
(401)737-4000

SOUTH CAROLINA
Aeronautics Commission
John W. Hamilton, Director
Drawer 1987
Columbia Metropolitan Airport
Columbia, SC 29202
(803)734-1700

SOUTH DAKOTA
Department of Transportation Services
Dean Schofield, Director of Planning
700 Broadway Ave. East
Pierre, SD 57501-2586
(605)773-3574

TENNESSEE
Office of Aeronautics, Department
of Transportation
David S. Fulton, Administrator
P.O. Box 17326
Nashville Metropolitan Airport
Nashville, TN 37217
(615)741-3208

TEXAS
Aeronautics Commission
C.A. Wilkins, Executive Director
P.O. Box 12607, Capitol Station
Austin, TX 78711
(512)476-9262

UTAH
Aeronautical Operations Division
Phillip N. Ashbaker, Director
135 North 2400 West
Salt Lake City, UT 84116
(801)823-2066

VERMONT
Agency of Transportation
Douglas E. Wheeler, Transportation
Aviation Administrator
120 State St.
Montpelier, VT 05603-0001
(802)828-2828

VIRGINIA
Department of Aviation
Kenneth A. Rowe, Director
P.O. Box 7716
4508 South Laburnum Ave.
Richmond, VA 23231
(804)786-1364

WASHINGTON
Division of Aeronautics
William H. Hamilton, Assistant
Secretary
8600 Perimeter Rd., Boeing Field
Seattle, WA 98108
(206)764-4131

WEST VIRGINIA
Office of Community and Industrial
Development
Fred Cutlip, Director of Community
Development
Building 6, B-553
State Capitol Complex
Charleston, WV 25305
(304)348-4010

WISCONSIN
Bureau of Aeronautics
Robert Kunkel, Acting Director
P.O. Box 7914
Madison, WI 53707
(608)266-3351

WYOMING
Aeronautics Commission
Richard Spaeth, Director
Cheyenne WY 82002-0090
(307)777-7481

Table 5-5. FAA REGIONAL OFFICES

ALASKAN REGION—ANCHORAGE
Governing Alaska and Aleutian Islands

Franklin L. Cunningham, Director
Anchorage Federal Office Building
701 C St.
P.O. Box 14
Anchorage, AK 99513
(907)271-5645

CENTRAL REGION—KANSAS CITY
Governing Iowa, Kansas, Missouri, Nebraska

Paul K. Bohr, Director
601 East 12 St.
Federal Building
Kansas City, MO 64106
(816)426-5626

EASTERN REGION—NEW YORK
Governing Delaware, District of Columbia, Maryland, New Jersey, New York, Pennsylvania, Virginia, West Virginia

Joseph M. Del Balzo, Director
John F. Kennedy International Airport
Fitzgerald Federal Building
Jamaica, NY 11430
(718)917-1005

EUROPE, AFRICA, AND MIDDLE EAST REGION—BRUSSELS, BELGIUM
Governing Europe, Africa, and the Middle East

Benjamin Demps Jr., Director
c/o American Embassy—FAA
APO, NY 09667-1011

GREAT LAKES REGION—CHICAGO
Governing Illinois, Indiana, Minnesota, Michigan, Ohio, Wisconsin, North and South Dakota

William H. Pollard, Director
O'Hare Lake Office Center
2300 East Devon Ave.
Des Plaines, IL 60018
(312)694-7000

NEW ENGLAND REGION—BOSTON
Governing Connecticut, Maine, Massachusetts, New Hampshire, Rhode Island, Vermont

Arlene B. Feldman, Administrator
12 New England Executive Park
Burlington, MA 01803
(617)273-7244

NORTHWEST MOUNTAIN REGION—SEATTLE
Governing Colorado, Idaho, Montana, Oregon, Utah, Washington, Wyoming

Wayne J. Barlow, Director
17900 Pacific Hwy. South
C-68966
Seattle, WA 98168
(206)431-2001

SOUTHERN REGION—ATLANTA
Governing Alabama, Florida, Georgia, Kentucky, Mississippi, North Carolina, South Carolina, Tennessee, Puerto Rico, U.S. Virgin Islands

Garland P. Castleberry, Director
P.O. Box 20636
Atlanta, GA 30320
(404)763-7222

SOUTHWEST REGION—FORT WORTH
Governing Arkansas, Louisiana, New Mexico, Oklahoma, Texas

C.R. (Tex) Melugin Jr., Director
4400 Blue Mound Rd.
Fort Worth, TX 76193
(817)624-5000

WESTERN-PACIFIC REGION—LOS ANGELES
Governing Arizona, California, Nevada, Hawaii, Pacific Ocean, and Asia

Homer C. McClure, Director
P.O. Box 92007
Worldway Postal Center
Los Angeles, CA 90009
(213)297-1427

6

Airport Management

C-5A Galaxy

This information is courtesy of FAA Aviation Career Series pamphlet GA-300-124-84.

AIRPORT MANAGEMENT AND OPERATIONS

Airports are usually operated by a director or manager that is responsible either to the private owners of the airport or to the local government authorities. The Airport Manager must be competent in public relations, economics, business management, civil engineering, personnel management, labor relations, and politics. The manager is involved in executive business decisions and could be responsible for:

- Making and enforcing airport rules and regulations.
- Planning and supervising maintenance and safety programs.
- Negotiating leases with airport tenants, such as airlines.
- Surveying future needs of the airport and making recommendations.
- Setting up the airport budget.
- Training and supervising employees.

Depending on the size of the airport, the manager might have one or more assistants such as an assistant manager, engineer, controller, personnel officer, maintenance superintendent, or supporting office workers such as secretaries, typists, and clerks.

If the manager is self-employed as a small airport operator, he or she probably also operates an aircraft repair station, sells aviation fuel, gives flight lessons and offers taxi or charter flights in addition to operating the airport.

Working Conditions. Working conditions vary greatly depending on the size of the airport. At a large airport, the manager works in an office usually located in the terminal building at the airport. Office hours are regular except in times of emergencies; travel might be required to negotiate leases with airline tenants or to confer with state and federal officials. If she or he operates a very small airport, this person might also work long hours giving flying lessons, making charter flights or supervising in the aircraft repair station. In such cases, much of the time is spent out of the office. In many cases, the Airport Manager is a part of the local government and is often involved in official meetings and community projects, especially those concerned with aviation.

Where the Jobs Are. Attended airports (those with a permanent, full-time work force) exist in every state. There is a higher concentration of these airports in California, Florida, Illinois, Indiana, Michigan, Missouri, Ohio, New York, Pennsylvania, and Texas. The scheduled airlines serve more than 600 airports, while the remaining attended airports are used primarily by general aviation aircraft.

Wages and Advancement. Unless the operator is a private owner and is self-employed, the job of an airport manager is not an entry-level job. Thus, it requires experience and training. An airport manager at a smaller airport can progress by moving to an assistant director's job at a larger airport and/or can also move upward to the position of commissioner of airports or to a state-level job concerned with state regulation at airports. Job opportunities often involve political activity and appointments frequently are made on that basis, especially if the job does not come under state or federal Civil Service regulations.

Dallas/Ft. Worth International Airport

Requirements to Enter the Job. Managers of airports with airline service are usually required to have a college degree in one of the following areas: airport management, business administration, or aeronautical or civil engineering. One study evaluated the importance of a number of educational areas in airport management. Besides a college degree, the study rates as ''very important'' a background in public relations, air transportation, business management, engineering, and personnel administration. The airport manager might be required to have had experience as an assistant at an airport. Managers of small airports qualify in some cases if they have only a high school diploma, but they must usually have a pilot's license and three to five year's experience in several kinds of jobs associated with airport services such as fixed base operator, superintendent of maintenance, or assistant to the airport manager.

The airport manager must be familiar with state and federal regulations (especially those pertaining to airports); zoning laws; environmental impact analysis; legal contracts; security; crash fire and rescue (CFR); and public relations. Airport managers must have strong leadership qualities, tact, initiative, good judgment, and an ability to get along with others.

Opportunities for Training. Numerous universities offer courses and degrees in airport administration, public administration, business administration, aeronautical or civil engineering, and flight training. Information on these schools can be obtained by writing to:

UNIVERSITY AVIATION ASSOCIATION
P.O. Box 2321
Auburn, Alabama 36830

Outlook for the Future. The economic contribution of airports has been estimated at several hundred billion dollars. Predictions of increases in the number of airline passengers, air cargo tonnage, and the production and use of general aviation aircraft over the next decade imply effects on airports. Existing airports will likely be enlarged as the general aviation fleet grows, but few more will come into existence due to environmental limitations. An increase in the number of fixed base operations (FBOs) is anticipated to service these additional aircraft; these FBOs will tend to be branches of existing nationwide FBO service companies rather than local private businesses. More smaller airports will come into existence, because towns without airports that are trying to attract new businesses will become aware of industry's increasing insistence on airport facilities near plant sites.

The present heavy air traffic at many larger airports has created landing and take-off delays. This has brought about the upgrading of smaller ''satellite'' airports to service general aviation fixed-wing aircraft and helicopters. This allows air travelers more options in traveling to the major airport for long-haul trips. This upgrading of smaller airport facilities will provide additional opportunities for airport managers and members of the supporting staffs.

Career Profile

Airport Manager

Name:		Danny L. Bruce
Career Field:		Airport Management
Position Held:		Director of Aviation: City of Dallas Dallas, Texas
Education:	**1974**	Masters Degree in Public Administration Southern Methodist University, Dallas, Texas
	1968	Bachelors Degree in Government University of Texas at Arlington
—		Regional Police Academy Graduate, Arlington, Texas
—		FAA-licensed private pilot
Experience:		
1980 - Present		*Director of Aviation*, Love Field, Dallas, Texas
1972 - 1980		*Assistant Director—City of Dallas*, Building Services Department
1969 - 1972		City of Dallas—Public Works Department
1965 - 1969		City of Dallas—Water Department
Professional Affiliations:		Airport Operators Council International (AOCI) American Association of Airport Executives (AAAE)

Job Responsibilities

In general, the airport manager is responsible for the overall operation and safety of the airport and its facilities. At large airports, the airport manager might have assistants who do research and give advice on business decisions. The size of the airport manager's staff depends greatly on the size and complexity of the airport. Typical business decisions made by the airport manager include setting up the airport budget, promoting the use of the airport, surveying future needs of the airport and making recommendations, making and enforcing airport rules and regulations, planning and supervising maintenance and safety programs, and negotiating leases with airport tenants.

The airport budget is carefully planned so that all necessary airport projects and plans can be funded. When the budget is tight, the airport manager must determine which improvements have priority and allot the funds accordingly.

Time and effort are also put into promoting the use of the airport. This is done through careful advertising and marketing strategies. The greater the demand for the airport services, the more profitable the airport becomes.

The airport manager should always be thinking of the future needs of the airport and making recommendations for these improvements. As the airport grows, the airport must expand to accommodate the increased flow of passengers, customers, and services.

To run smoothly, there must be a given set of rules and regulations governing airport operations. Some rules and regulations are governed by the FAA and are outlined in various Federal Aviation Regulations (FARs). The airport manager helps to develop other procedures that are not covered under FARs but are necessary for smooth, efficient airport operations.

The airport manager is responsible for planning and supervising maintenance of the airport and its facilities. This includes maintaining taxiways, runways, parking garages, lighting, landscaping, roads, buildings, and other facilities. If the airport is ever judged to be in an unsafe condition, the airport could be closed until the repairs are accomplished. This could mean a substantial loss in time and money for the airport, so the job of keeping the airport maintained is an extremely important one. Safety programs are also followed to ensure that customers of the airport can use the facilities with a maximum degree of safety. When a crisis does occur, the airport manager must be able to deal with the media and other investigating agencies in a calm and professional manner. This requires a certain amount of public relations and communication skills.

Finally, the airport manager negotiates leases with airport tenants. Within the airport community, many different businesses and services operate on a daily basis. Such organizations include airline operators, aircraft repair/leasing/rental and charter operators, fixed-base operators (FBOs), restaurants, gift shops and concessions, auto parking, crash and rescue operations, medical operations, security police, ice and snow removal operations, and various other organizations and services. In short, the airport manager acts as the landlord for these business operations. The airport manager usually works directly for the airport owner. This could be a private owner or the local government authorities. In my case, the owner of Love Field is the city of Dallas.

That's a look at the internal airport responsibilities. The airport manager also has various extra airport considerations as well. He or she must also interface with the airport's neighboring communities and ensure that public relations are healthy. Topics such as increased road traffic, air pollution, and noise pollution are subjects that can degrade airport-community relations.

Running an airport is an enterprise operation like any other business. The airport manager must have a good grip on the many different roles of such a diverse occupation. He or she must be able to coordinate the efforts of the FAA, the public, the other operating agencies, and the airport at hand.

Benefits and Drawbacks of the Job

Working as an airport manager, in my opinion, is one of the most exciting aviation careers. Managers usually experience a variety of work and interface with many different people. The job can be very challenging and full of energy. There is also a certain amount of satisfaction in seeing the airport operate efficiently.

This type of job requires a lot of personal involvement, which some might find very taxing. Getting the job done can require a lot of long hours in the office and being on call the rest of the day and night. If an incident or accident does occur on airport grounds at 2:00 a.m., the airport manager might be required to attend the late-night investigation and help to resolve the problem. The investigation could end early in the morning, just in time for the start of another working day . . . it is all part of the job.

When working with the FAA, the public, and the operating staff, it can often be difficult to please all the people all the time. The airport manager has the unique task of finding the best solution for the given situation and implementing it. It's a big job with big responsibilities and big rewards.

Recommendations to Job Hunters

Airport management jobs require a good deal of business sense. To prepare for such a job, the applicant should pursue a business-related education. Because a good deal of the job requires analyzing proposals and making business recommendations, a certain amount of analytical ability is also required. The manager must decide what deals are good for the airport and which are not worth pursuing.

It is also important to have a good working knowledge of aviation. It would be hard to make decisions on aviation-related subjects if you did not understand the day-to-day operations and problems associated with them. Being a pilot would also be very beneficial, although not necessary, because among other things, it helps to understand the "lingo" associated with the aviation community.

Work experience is an important asset. Two good sources of aviation experience are working with an airline or some type of air carrier operation, or working in the military in an aviation-related career. This type of experience would give the applicant a well-rounded background and would help to prepare him for a management job.

Airport managers come from a variety of different backgrounds. They all, however, have a few things in common. They are business-minded people with an appreciation of aviation and a well-rounded background and personality.

ASSISTANT AIRPORT MANAGER

The assistant helps the manager carry out administrative responsibilities and might be in charge of maintenance employees, airport equipment, airport tenant relations, or any of the other kinds of work associated with an airport. The salary varies according to the size of the airport and the number of duties. Therefore, requirements for the job of assistant manager range from a high school diploma to a college degree in business or engineering. The larger metropolitan airports typically require three to seven years of experience at airports served by several airlines.

ENGINEER

Airport engineers plan improvements and expansion of the airport, check on plans submitted by architects and contractors, oversee construction, handle real estate and zoning problems, and might direct maintenance of runways, taxiways, hangars, terminal buildings and grounds. Engineers are employed mostly by larger airports. A degree in civil engineering is normally preferred, plus three to seven years of experience.

AIRPORT PLANNER OR CONSULTANT

Perhaps one of the most critical jobs at the airport belongs to the highly skilled individuals who develop the airport master plan. Some of the responsibilities of this position include planning runways, taxiways, terminal buildings, parking areas, FAA facilities, and other structures necessary for the day-to-day operation of the airport. Airport planners must also consider how the airport will impact the surrounding communities. Environmental considerations such as pollution, noise, and traffic congestion are items that must be planned well in advance to eliminate any unfriendly relations.

The job does not end once the airport is built. Airport planners must also monitor the need for growth and expansion. This is done by forecasting the usage of the airport and setting up short- and long-range airport goals.

To meet this need, the airport might have a developing and planning group on their staff. This is typical at many of the larger airports in this country. Many of the smaller airports can not justify having a full-time planning and developing staff, so they normally hire a consulting firm to do the necessary study. The consulting firm would then submit a written document with their recommendations and conclusions to the airport committee.

According to Dana Ryan, a senior planner at Dallas/Fort Worth International Airport, their airport planning staff consists of approximately 100 people. The planners broken down into four basic groups: planning, design, construction, and contracts. These four groups work together to incorporate any changes necessary to meet the growing traffic demands at the airport. Present-day statistics show that approximately 40 to 50 thousand vehicles per day travel over the airport grounds. Airport planners must consider where these vehicles will park, how the traffic will flow, and how to accommodate the individual needs of each airport patron.

The planning group is responsible for development and planning of the facilities, ground transportation, and any other special projects that arise. The design group is responsible for the architectural layout and design of the facilities. The construction group is responsible for the actual building of the facilities (in many cases, the airport contracts outside firms to perform the construction). Finally, the contracts group determines whether the funds to entertain such a project exist. At present, it is estimated that Dallas/Fort Worth Airport spends approximately 60 to 100 million dollars a year in capital improvements. Given this, it becomes obvious why the airport planner is such a necessary individual.

Qualifications for such a position vary depending on the airport and the position offered. Many airports are looking for degreed individuals with some work experience. Desired academics include Civil Engineering, Architecture, Management, or any other relevant field of study.

AIRPORT LOGISTICS

The remaining employees on the airport manager's staff perform the necessary duties of supporting the main airport operations: controller, secretary, typist, etc.

AIRPORT SERVICEPERSON

These employees work under the direction of the airport manager or engineer and do one or more of the following jobs:

- Cut grass on airport grounds and maintain shrubbery.
- Operate snow-removal equipment.
- Service runway lights.
- Maintain electrical service on airport grounds, paint, and do general carpentry work on smaller repair jobs.

Large airports employ workers who specialize in one of the foregoing jobs, for example, airport electrician. Many trades and crafts are needed at the airport, although small airports usually contract for required maintenance. Training for such jobs, qualifications, wages, opportunity for advancement, and union agreements are the same as for other workers in the trades and crafts.

SAFETY EMPLOYEES

Most airports with airline service must maintain crash fire-fighting and rescue (CFR) equipment. This provides employment for small numbers of trained firefighters and rescue workers, some of whom may be trained as emergency medical technicians (EMTs) or paramedics. Typically, airport firefighters develop the skills of both aircraft and building or structural firefighting.

TERMINAL CONCESSIONAIRES

Airport terminals provide services for air passengers such as restaurants, news stands, gift and book shops, car rental agencies and skycap baggage service. (Only a few airlines employ skycaps; most leave this service to terminal concessions.) Workers in the airport flight kitchens cater to airlines that do not have their own flight kitchens. While not on the airport manager's staff, workers in the foregoing concessions are mentioned only because they have a place in the total employment picture of the airport.

FIXED-BASE OPERATOR (FBO)

A fixed-base operator is a retail firm that sells general aviation products or services at an airport. The FBO might employ one or two people or have 100 workers. One or more of the following services are offered: aircraft airframe, engine, and/or instrument repairs; flight training; air taxi service and charter flights; aircraft sales; aircraft fuel and parts; and aircraft exterior and/or interior modification.

The FBO employs aviation mechanics, flight instructors, and aircraft salespersons, depending on the size and scope of the operation. The FBO might also carry and service a small aviation mechanic's training operation. The FBO might even arrange for ground transportation and overnight accommodations for general aviation pilots and their passengers.

The FBOs place of business could be a small hangar or shop with adjoining office and perhaps a pilot's lounge, or it could be an elaborate series of hangars, shops, offices, classrooms, and showrooms. The hours are determined by the amount of time the FBO wants to devote to the business.

As the FBO is essentially an entrepreneur, the opportunities for increased business and income depend on the operator's initiative and ability to keep up with changes in aircraft, aircraft equipment, and services. The variety of activities in which an FBO can be involved offer some assurance of stability of income.

It is difficult to determine just what the requirements are to become a fixed-base operator. Certainly an interest in aviation and training as an aviation mechanic are basic. A pilot's license is not essential, but it would provide both a greater understanding of the many functions of an FBO and provide contact with pilots who might patronize the business. With a commercial pilot's license, the operator could supplement the FBO income with air taxi and charter flights, and with a flight instructor's license, the FBO could give flight lessons. Training in business administration is also helpful in setting up an efficient business operation and is proving to be essential to success in times of reduced aviation activity brought on by the recent downturn in the national economy.

The professional fixed-base operator has a bright future. Today's general aviation active fleet numbers to about 220,000 aircraft. By 1990, this figure is expected to increase to about 310,000. All these aircraft will need fuel, parts, accessories, regular maintenance, overhauls and pilots to fly them. Aircraft servicing and maintenance, flight instruction, and fuel are three services offered by most FBOs. They will continue to be offered in increasing volume as the general aviation fleet increases in numbers. Airports that may be unattended at present will gradually offer various services to the general aviation pilot. Airports now having FBOs as tenants could add more FBO facilities. Predictions for opportunities as a fixed-base operator are optimistic over the next 15 years with improvement in the economy, especially if the FBO can attract customers with an efficient, orderly service.

LINEPERSON

The fixed-base operator employs linepersons or ramp servicepersons who meet arriving aircraft, guide them to parking spots, assist pilots to tie down their aircraft, and perform the important duties of serving general aviation and airline customers. These duties include fueling and servicing aircraft. One important function is reporting to the owner signs of incipient trouble with aircraft, such as oil leaks.

Linepersons are usually young people who are interested in aviation and begin their careers by building up experience with aircraft under the guidance of a fixed-base operator. They are usually paid an hourly rate and often work part time after school hours, on weekends, or during the summer. With their earnings, they often fly or take up aviation maintenance. The lineperson's job is an important basic career development step and leads to many aviation careers, especially those associated with airport administration, fixed-base operator, aviation mechanic, professional pilot, or air traffic controller.

7
Job-Hunting Strategy

Navy T-44A

This chapter contains a few pointers on how to find the occupation you're looking for once you have the expected education and/or experience.

EMPLOYMENT AGENCIES

Using an employment agency can be a distinct benefit to the job hunter. Before using one, however, an understanding of what they are and how they work is necessary.

Most employment agencies operate as businesses, not as counseling services. They are not in business to help you decide what interests you should pursue. The employment agent is a knowledgeable individual who follows the trends in the employment market. They know what organizations are hiring, what skills those organizations are looking for, and who to contact for details. This can be a distinct advantage to the job hunter in gaining maximum exposure of his or her resume for a minimum amount of effort.

When seeking a job agency, for best results, look for those who specialize in your area of interest. Specialized agencies usually track job openings on a continual basis so they are way ahead of the program.

Another advantage of using an employment agency is that most agencies have nationwide affiliates that can aid in job hunting at a distance. Each agency has contacts in its particular area and is more familiar with the day-to-day operations of local employers. Using the employment agency gives the job hunter a nationwide network of job exposure.

Like most businesses, the employment agency does not work for free. Someone will compensate the agency for its efforts when a job is placed through their office. Agencies receive a service fee based on the applicant's starting salary. For most professional positions, the employer pays the service fee. These are known as "fee paid" positions; the agency's services are provided to the applicant at no expense. Before signing any contract with an agency, be sure you know who is responsible for the fee. If you are interested in only "fee paid" positions, be sure your agent is aware of your intent. When the fee is not "fee paid," the applicant will be responsible for just compensation to the agency.

Note: If you, as an applicant accept a job through the agency and do not report to the job, of if you leave on your own accord while in the "probationary period" (usually a few months after hire), then you might be liable for the fee paid by the employer. Know the terms of your contract before signing anything.

Employment agencies provide valuable information and exposure for the applicant. They can help answer questions, give the applicant an honest evaluation of the job market, and also provide advice on resumé writing and interview techniques. You can switch agencies if you're not pleased with the performance of the organization, and you can use more than one agency if desired. The agency is there to help you find the job you're looking for, so take advantage of all of the service they offer. Employment agencies are listed in various trade magazines and newspapers. Refer to the magazine and newspaper listing in Table 7-1. Further, a complete listing of private Employment Agencies can be obtained by writing to

NATIONAL ASSOCIATION OF PERSONNEL CONSULTANTS
1432 Duke Street
Alexandria, VA 22314

PUBLICATIONS

The following publications listing of Table 7-1 contains newspapers and other publications that carry extensive employment advertising. Short-term (three- or six-month) subscriptions for the Sunday edition are convenient for the out-of-town job hunter.

ORGANIZATIONS

Aviation organizations, unions, and technical societies are organizations that harbor a wealth of information in each field of expertise. These organizations can be a great asset to the job hunter. A quick letter or phone call can gain valuable advice and guidance to further the job hunting venture. Many of these organizations are listed in Table 7-2.

Table 7-1. AVIATION PUBLICATIONS

NEWSPAPERS

Atlanta-Journal-Constitution
P.O. Box 4689
Atlanta, GA 30302

Boston Globe
135 Morrissey Blvd.
Boston, MA 02107

Chicago Tribune
435 North Michigan Ave.
Chicago, IL 60611

Dallas Morning News
Communications Center
Dallas, TX 75265

Denver Post
P.O. Box 1709
Denver, CO 80201

Houston Post
4747 Southwest Freeway
Houston, TX 77001

Los Angeles Times
Times Mirror Square
Los Angeles, CA 90053

Minneapolis Star & Tribune
425 Portland Ave.
Minneapolis, MN 55488

New York Times
229 West 43rd St.
New York, NY 10036

Philadelphia Inquirer
400 North Broad St.
Phildelphia, PA 19101

St. Louis Post-Dispatch
900 North Tucker Blvd.
St. Louis, MO 63101

San Francisco Chronicle
901 Mission St.
San Francisco, CA 94103

Seattle Post-Intelligencer
Sixth and Wall St.
Seattle, WA 98121

Wall Street Journal, Eastern Edition
Dow Jones & Company, Inc.
22 Cortlandt St.
New York, NY 10007

Wall Street Journal, Midwest Edition
Dow Jones & Company, Inc.
200 West Monroe St.
Chicago, IL 60606

Wall Street Journal, Southwest Edition
Dow Jones & Company, Inc.
1233 Regal Row
Dallas, TX 75247

Wall Street Journal, Western Edition
Dow Jones & Company, Inc.
220 Battery St.
San Francisco, CA 94111

Washington Post
1150 15th St., N.W.
Washington, DC 20071

MAGAZINES AND TRADE PUBLICATIONS

AG Pilot International
B 16 - Aeronautic Publishers, Inc.
Drawer R
Walla Walla, WA 99362

Agricultural Aviation
National Agricultural Aviation Assoc.
115 D St., S.E.
Washington, DC 20003

Air Cargo World
Communication Channels, Inc.
6255 Barfield Rd.
Atlanta, GA 30328

Air Force Times
Army Times Publishing Co.
Springfield, VA 22159

Air Line Pilot
Air Line Pilots Assoc.
535 Herndon Pkwy.
Box 1169
Herndon, VA 22070

Airline and Travel Food
International Publishing Co.
of America
665 La Villa Dr.
Miami Springs, FL 33166

Airline Executive
Communication Channels, Inc.
6255 Barfield Rd.
Atlanta, GA 30328

Airline Quarterly
Challenge Publications
7950 Deering Ave.
Canoga Park, CA 91304

Air Market News
General Publications Co.
7272 Cradlerock Way
Columbia, MD 21045

Air Progress
Challenge Publications, Inc.
7950 Deering Ave.
Canoga Park, CA 91304

Air Transport World
Penton-IPC
Reinhold Publishing Division
600 Summer St., Box 1361
Stamford, CT 06904

Airport Services Management
Lakewood Publications, Inc.
731 Hennepin Ave.
Minneapolis, MN 55403

AOPA Pilot
421 Aviation Way
Frederick, MD 21701

Aviation Equipment Maintenance
Irving-Cloud Publishing Co.
7300 Cicero Ave.
Lincolnwood, IL 60646

Aviation International News
190 Godwin Ave.
Midland Park, NJ 07432

Aviation Magazine
Data Publications, Inc.
Box 2231
Danbury, CT 06810-2231

Aviation Week and Space Technology
P.O. Box 1505
Neptune, NJ 07754-1505

Aviation/Space Magazine
Aerospace Education Assoc.
1810 Michael Faraday Dr.
No. 101
Reston, VA 22090-5391

Avionics
Box 5100
Westport, CT 06881

Avionics News Magazine
Aircraft Electronics Assoc.
Box 1981
Independence, MO 64055

Business & Commercial Aviation
P.O. Box 5850
Cherry Hill, NJ 08034

Commercial Space
McGraw-Hill Publications Co.
1221 Ave. of the Americas
New York, NY 10020

Commuter Air
Communication Channels Inc.
6255 Barfield Rd.
Atlanta, GA 30328

Electronic Design
P.O. Box 1036
Southeastern, PA 19398

EDN
2700 St. Paul St.
Denver, CO 80206

Electronic Products
645 Stuart Ave.
Garden City, NY 11530

FAA General Aviation News
Supt. of Documents
Washington, DC 20402

Flight International
Business Press International LTD
Quadrant House,
The Quadrant
Sutton, Surrey
SM2 5AS
England

Flightline
Pilots International Assoc. Inc.
4000 Olson Memorial Hwy.
Minneapolis, MN 55422

Flying
P.O. Box 51377
Boulder, CO 80321-1377

Helicopter International Magazine
Avia Press Associates
75 Elm Tree Rd.
Locking, Weston-Super Mare
Avon BS24 8 EL
England

IEEE Spectrum
445 Hoes Lane
Piscataway, NJ 08854

IFR Magazine
IFR Publishing Co, Inc.
12940 W. National Ave.
Box 244
New Berlin, WI 53151

International Aviation Mechanics Journal
Aviation Maintenance Publishers Inc.
Box 36
Riverton, WY 82501-0036

Jet Cargo News
Hagall Publishing Co.
Box 920952
No. 398
Houston, TX 77292-0952

Journal of Air Traffic Control
Air Traffic Control Assoc.
2020 N. 14th, St. Suite 410
Arlington, VA 22201

Journal of Guidance Control & Dynamics
Box 11312
Church St. Station
New York, NY 10249

Midwest Flyer Magazine
Flyer Publications
Box 199
Oregon, WI 53576-0199

NASA Activities
Supt. of Documents
U.S. Gov't. Printing Office
Washington, DC 20402

Naval Aviation News
Supt. of Documents
Washington, DC 20402

Navy Times
Army Times Publishing Co.
Springfield, VA 22159

PAMA News
Professional Aviation
Maintenance Assoc.
Box 248
St. Ann, MO 63074

Passenger & In-Flight Service
International Publishing
Co. of America
665 La Villa Dr.
Miami Springs, FL 33166

Pilot News
Suburban Pilot, Inc.
5320 N. Jackson Ave.
Kansas City, MO 64119

Plane & Pilot
Werner and Werner Corp.
16200 Ventura Blvd. Suite 201
Encino, CA 91436

Plane & Pilot
Werner and Werner Corp.
16200 Ventura Blvd., Suite 201
Encino, CA 91436

Private Pilot
Box 6040
Mission Viejo, CA 92690

Professional Pilot Magazine
Queensmith Communications Corp.
West Building
Washington National Airport
Washington, DC 20001

Rotor & Wing International
PJS Publications Inc.
News Plaza, Box 1790
Peoria, IL 61656

Space Age Times
U.S. Space Education Assoc.
News Operations Division
746 Turnpike Rd.
Communications Office
Elizabethtown, PA 17022

Sport Aviation
Experimental Aircraft Assoc. Inc.
Box 3086
Oshkosh, WI 54903

TAC Attack
U.S. Dept. of the Air Force
Tactical Air Command
Langley AFB, VA 23665-5001

Test Engineering & Management
3756 Grand Ave.
Suite 205
Oakland, CA 94610

United States Army Aviation Digest
Supt. of Documents
Washington, DC 20402

U.S. News and World Report
2400 N St. NW
Washington, DC 20037

Vertica
Pergamon Press Inc.
Journals Division
Maxwell House
Fairview Park
Elmsford, NY 10523

Vertiflite
American Helicopter Society Inc.
217 N. Washington St.
Alexandria, VA 22314

World Air News
TCE Publications LTD
Eastgate House
Four Forks, Spaxton
Nr. Bridgewater
Somerset
England

Table 7-2. Aviation Organizations

Academy of Model Aeronautics
1810 Samuel Morse Dr.
Reston, VA22090
(703)435-0750

Aero Club of Pittsburgh
Erik Wagner, President
207 Dormont Village
2961 West Liberty Ave.
Pittsburgh, PA 15216
(412)341-5090

Aerobatic Club of America
3001 North Lakeridge Trail
Boulder, CO 80302
(303)449-5291

Aeronautical Radio, Inc. (ARINC)
2551 Riva Rd.
Annapolis, MD 21401
(301)266-4000

AERONCA AVIATORS CLUB
511 Terrace Lake Rd.
Columbus, IN 47201
(812)342-6878

AERONCA LOVER'S CLUB
Buzz Wagner
P.O. Box 3
401 1st St. East
Clark, SD 57225
(605)532-3862

AERONCA SEDAN CLUB
Richard Welsh
2311 E. Lake Sammamish Place, SE
Issaquah, WA 98027

AEROSPACE EDUCATION ASSOCIATION
1810 Michael Faraday Dr., Suite 101
Reston, VA 22090
(703)435-4449

AEROSPACE ELECTRICAL SOCIETY
P.O. Box 24BB3
Village Station
Los Angeles, CA 90024

AEROSPACE INDUSTRIES ASSOCIATION OF AMERICA, INC.
1250 I St., N.W., Suite 1100
Washington, DC 20005
(202)371-8400

AEROSPACE MEDICAL ASSOCIATION
R.R. Hessberg, M.D.
320 South Henry St.
Alexandria, VA 22314-3524
(703)739-2240

AEROSTAR OWNERS ASSOCIATION
Karen Griggs
500 E St., S.W., Suite 930
Washington, DC 20024
(202)863-1000

AIRBORNE LAW ENFORCEMENT ASSOCIATION, INC.
Major R. R. Raffensberger
Baltimore Police Department
601 East Fayette St.
Baltimore, MD 21202
(301)396-2431

AIRCHIVE ASSOCIATION
Job C. Conger IV, Executive Director
521 South Glenwood
Springfield, IL 62704
(217)789-9754

AIRCRAFT ELECTRONICS ASSOCIATION
Monte Mitchell, Executive Director
P.O. Box 1981
Independence, MO 64055
(816)373-6565

AIRCRAFT OWNERS AND PILOTS ASSOCIATION (AOPA)
421 Aviation Way
Frederick, MD 21701
(301)695-2000

AIR FORCE ASSOCIATION
Charles L. Donnelly, Jr.
Executive Director
1501 Lee Hwy.
Arlington, VA 22209-1198
(703)247-5800

AIRLINE INDUSTRIAL RELATIONS CONFERENCE
Suite 250, 1920 N St., N.W.
Washington, DC 20036
(202)861-7550

AIR LINE PILOTS ASSOCIATION (ALPA)
535 Herndon Parkway
P.O. Box 1169
Herndon, VA 22070
(703)689-2270

AIRLINE SERVICES ASSOCIATION
1101 Connecticut Ave. N.W.
Suite 700
Washington, DC 20036

AIRPORT OPERATORS COUNCIL
Suite 800, 1220 19th St. N.W.
Washington, DC 20036
(202)293-8500

AIR TRAFFIC CONTROL ASSOCIATION
2020 North 14th St., Suite 410
Arlington, VA 22201
(703)522-5717

AIR TRANSPORT ASSOCIATION OF AMERICA (ATA)
1709 New York Ave., N.W.
Washington, DC 20006
(202)626-4000

ALLIED PILOTS ASSOCIATION
2214 Paddock Way Dr., Suite 900
Grand Prairie, TX 75050
(214)988-3188

AMERICAN AIR RACING SOCIETY
Rudy Profant
4060 West 158th St.
Cleveland, OH 44135

AMERICAN ASSOCIATION OF AIRPORT EXECUTIVES
4224 King St.
Alexandria, VA 22302
(703)824-0500

AMERICAN AVIATION HISTORICAL SOCIETY
2333 Otis St.
Santa Ana, CA 92704
(714)549-4818

AMERICAN BONANZA SOCIETY
Cliff R. Sones
Mid-Continent Airport
1922 Midfield Rd.
P.O. Box 12888
Wichita, KS 67277
(316)945-6913

AMERICAN ELECTRONICS ASSOCIATION
P.O. Box 10045
2670 Hanover St.
Palo Alto, CA 94303
(415)857-9300

AMERICAN HELICOPTER SOCIETY, INC.
217 North Washington St.
Alexandria, VA 22314-2538
(703)684-6777
Fax: (703)739-9279

AMERICAN INSTITUTE OF AERONAUTICS AND ASTRONAUTICS
Cort Durocher, Executive Director
370 L'Enfant Promenade, S.W.
Washington, DC 20024

AMERICAN NAVION SOCIETY
P.O. Box 1175
Municipal Airport
Banning, CA 92220
(714)849-2213

AMERICAN TIGER CLUB
Frank Price, President
Route 1, P.O. Box 419
Moody, TX 76557
(817)853-2008

AMERICAN YANKEE ASSOCIATION
Stewart Wilson
3232 Western Dr.
Cameron Park, CA 95682
(916)676-2022

ANIMAL AIR TRANSPORTATION ASSOCIATION, INC.
Box 441110
Fort Washington, MD 20744
(301)292-1970
Telex: 4997385 AATA
Fax: (301)292-1787

ANTIQUE AIRPLANE ASSOCIATION
Route 2, Box 172
Ottumwa, IA 52501
(515)938-2773

ARMY AVIATION ASSOCIATION OF AMERICA
49 Richmondville Ave.
Westport, CT 06880-2000
(203)226-8184

ASSOCIATION OF AVIATION PSYCHOLOGISTS
Thomas M. McCloy, Ph.D.
3237 Teardrop Circle
Colorado Springs, CO 80917

AVIATION DISTRIBUTORS AND MANUFACTURERS ASSOCIATION (ADMA)
1900 Arch St.
Philadelphia, PA 19103
(213)564-3484

Aviation Maintenance Foundation
Richard Kost, Executive Director
P.O. Box 2826
Redmond, WA 98073
(206)823-0633

Association of Naval Aviation
Suite 200, 5205 Leesburg Pike
Falls Church, VA 22041
(703)998-7733

Aviation/Space Writers Association
Madeline M. Field, Executive Director
17 South High St., Suite 1200
Columbus, OH 43215
(614)221-1900
Fax: (614)221-1989

Balloon Federation of America
P.O. Box 400
Indianola, IA 50125
(515)961-8809

Bellanca-Champion Club
Pam Foard
1820 North 166th St.
Brookfield, WI 53005
(414)784-0318

Bird Airplane Club
Jeannie Hill
P.O. Box 328
Harvard, IL 60033
(815)943-7205

Bucker Club
John Bergeson
6438 West Millbrook Rd.
Remus, MI 49340
(517)561-2393

Buckeye Pietenpol Association
Frank S. Pavliga, Newsletter Editor
2800 South Turner Rd.
Canfield, OH 44406
(216)792-6269

California Wheelchair Aviators
Bill Blackwood
1117 Rising Hill Way
Escondido, CA 92025
(619)746-5018

Cardinal Club
Beth Harrison, President
1701 Saint Andrew's Dr.
Lawrence, KS 66046

Cessna Airmaster Club
Gar Williams
9 South 135 Aero Dr.
Naperville, IL 60565

Cessna Owner Organization
Garry L. Dollahite, President
P.O. Box 75068
3447 Lorna Lane
Birmingham, AL 35253
(205)822-8035 or (800)247-8360

Cessna Pilots Association
John Frank, Executive Director
Wichita Mid-Continent Airport
2120 Airport Rd.
P.O. Box 12948
Wichita, KS 67277
(316)946-4777

Cessna 150/152 Club
Skip Carden, Executive Director
P.O. Box 15388
Durham, NC 27704
(919)471-9492

Cessna 172/152 Straight Tail Owners
Ernie Colbert
9801 South A1A, Lot 652-2
Jensen Beach, FL 33457

Cherokee Pilots' Association
Terry L. Rogers, Executive Director
P.O. 716
Safety Harbor, FL 34695
(813)791-3255

China-Burma-India Hump Pilots Association
Dr. Herbert O. Fisher, Director of Public Relations and Press
628 Mountain Rd., Smoke Rise
Kinnelon, NJ 07405
(201)838-2040

CHRISTIAN PILOTS ASSOCIATION, INC.
Howard Payne, President
P.O. Box 603
West Covina, CA 91793-0603
(818)962-7591

CIVIL AIR PATROL
National Headquarters
National Administrator
Building 714
Maxwell AFB, AL 36112-5572
(205)293-6019

CIVIL AVIATION MEDICAL ASSOCIATION
Albert Carriagere, Business Counsel
755 Bank Lane
Lake Forest, IL 60045
(312)234-6330

COMBAT PILOTS ASSOCIATION
Paul Conger, President
Box 1831
Glendora, CA 91740

COMMANDER FLYING ASSOCIATION
899 West Foothill Blvd., Suite E
Monrovia, CA 91016-1938
(818)359-1040

COMMUTER AIRLINE ASSOCIATION OF AMERICA
1101 Connecticut Ave. N.W.
Suite 700
Washington, DC 20036
(202)857-1170

CONFEDERATE AIR FORCE
P.O. Box CAF
Harlingen, TX 78551
(512)425-1057

CONTINENTAL LUSCOMBE ASSOCIATION
Loren Bump, President
5736 Esmar Rd.
Ceres, CA 95307
(209)537-9934

CORBEN CLUB
Robert L. Taylor, Editor
P.O. Box 127
Blakesburg, IA 52536
(515)938-2773

CUB CLUB
John Bergeson
P.O. Box 2002
Mount Pleasant, MI 48858
(517)561-2393

CULVER CLUB
Larry Low, Chairman
60 Skywood Way
Woodside, CA 94062
(415)851-0204

CULVER PQ-14 ASSOCIATION
Ted Heineman, Editor
29621 Kensington Dr.
Laguna Niguel, CA 92677
(714)831-0173

DART CLUB
Lloyd Washburn
3948 Washburn Dr.
Port Clinton, OH 43452

DEHAVILLAND MOTH CLUB
Gerry Schwam, Chairman
1021 Serpentine Lane
Wyncote, PA 19095
(215)635-7000 or (215)866-8283

DEHAVILLAND MOTH CLUB OF CANADA
R. deHavilland Ted Leonard
Founder-Director
305 Old Homestead Rd.
Keswick, Ontario, Canada L4P 1E6
(416)476-4225

EARLY BIRDS OF AVIATION, INC.
J. Emery Stromberg, Secretary
2233 East Oakmont Ave.
Orange, CA 92667
(714)639-1839

EASTERN 190/195 ASSOCIATION
Cliff Crabs
25575 Butternut
North Olmsted, OH 44070
(216)777-4025

EASTERN REGION/UNITED STATES AIR RACING ASSOCIATION, (ER/USARA)
26726 Henry Rd.
Bay Village, OH 44140
(216)961-9010

ELECTRONIC TECHNICIANS ASSOCIATION INTERNATIONAL
Route 3, Box 564
Greencastle, IN 46135

ERCOUPE OWNERS CLUB
Skip Carden, Executive Director
P.O. Box 15388
Durham, NC 27704
(919)471-9492

EXPERIMENTAL AIRCRAFT ASSOCIATION, INC., (EAA)
3000 Poberezny
Oshkosh, WI 54903-3086
(414)426-4800

FAIRCHILD CLUB
John W. Berendt, President
7645 Echo Point Rd.
Cannon Falls, MN 55009
(507)263-2414

FLEET CLUB
George G. Gregory, President
4880 Duguid Rd.
Manlius, NY 13104
(315)682-6380

FLIGHT ENGINEERS INTERNATIONAL ASSOCIATION
905 16th St. N.W.
Washington, DC 20006
(202)347-4511

FLIGHT SAFETY FOUNDATION, INC.
John H. Enders, President
5510 Columbia Pike
Arlington, VA 22204-3194
(703)820-2777
Telex: 901176, Fax: (703)820-9399

FLORIDA AERO CLUB
Karen Castino
2808 North 34th Ave.
Hollywood, FL 33021
(305)987-4286

FLYING APACHE ASSOCIATION
John and Carol Lumley
1 South Dillon Lane
Villa Park, IL 60181
(312)627-8027

FLYING ARCHITECTS ASSOCIATION, INC.
Harold J. Westin
Architects-Engineers, P.A.
2504 Manitou Island
White Bear Lake, MN 55110
(612)429-4385

FLYING CHIROPRACTORS ASSOCIATION
Dr. W.J. Quinlan
7301 Hasbrook Ave.
Philadelphia, PA 19111
(215)722-7200

FLYING DENTISTS ASSOCIATION
Ted Rigelman
12 East St.
Parkville, MO 64152
(814)741-0112

FLYING ENGINEERS INTERNATIONAL
8506 Louis Dr., S.E.
Huntsville, AL 35802

FLYING FUNERAL DIRECTORS OF AMERICA
Phillip North Schmidt
10980 Reading Rd.
Sharonville, OH 45241
(513)948-1113

FLYING PHYSICIANS ASSOCIATION
Don Drake
P.O. Box 17841
Kansas City, MO 64134
(816)763-9336

FLYING VETERINARIANS ASSOCIATION
Richard J. Rossman, D.V.M.
330 Waukegan Rd.
Glenview, IL 60025
(312)729-5200

FUNK AIRCRAFT OWNERS ASSOCIATION
G. Dale Beach, Editor-Treasurer
1621 Dreher St.
Sacramento, CA 95814
(916)443-7604

FUTURE AVIATION PROFESSIONALS OF AMERICA
4291-J Memorial Dr.
Decatur, GA 30032
1-800-JET-JOBS

GENERAL AVIATION MANUFACTURERS ASSOCIATION, (GAMA)
1400 K St., N.W.
Suite 801
Washington, DC 20005
(202)393-1500

GREAT LAKES CLUB
Robert L. Taylor, Editor
P.O. Box 127
Blakesburg, IA 52536
(515)938-2773

HATZ CLUB
Robert L. Taylor, Editor
P.O. Box 127
Blakesburg, IA 52536
(515)938-2773

HEATH PARASOL CLUB
William Schlapman
6431 Paulson Rd.
Winneconne, WI 54986
(414)582-4454

HELICOPTER ASSOCIATION INTERNATIONAL
Frank L. Jensen Jr., President
1619 Duke St.
Alexandria, VA 22314
(703)683-4646

HELICOPTER FOUNDATION INTERNATIONAL
Robert C. Kerner, Executive Director
1619 Duke St.
(703)683-4646

HELICOPTER CLUB OF AMERICA
John F. Zugschwert, President
217 North Washington St.
Alexandria, VA 22314
(703)684-6777

INDEPENDENT FEDERATION OF FLIGHT ATTENDANTS
630 Third Ave.
New York, NY 10017

INTERNATIONAL 180/185 CLUB
Charles Bombardier
4539 North 49th Ave.
Phoenix, AZ 85031
(602)846-6236

INTERNATIONAL 195 CLUB
Dwight M. Ewing, President
P.O. Box 737
Merced, CA 93544
(209)722-6283

INTERNATIONAL ACADEMY OF ASTRONAUTICS
3-5 Rue Mario-Nikis, F-75015
Paris, France
1-456-74966

INTERNATIONAL AEROBATIC CLUB
Wittman Field
P.O. Box 3086
Oshkosh, WI 54903-3086
(414)426-4800

INTERNATIONAL AIRLINE PASSENGERS ASSOCIATION
P.O. Box 660074
Dallas, TX 75266
(214)404-9980

INTERNATIONAL AIR TRANSPORT ASSOCIATION
IATA Building, 2000 Peel St.
Montreal, Quebec, Canada H3A 2R4
(514)844-6311

INTERNATIONAL ASSOCIATION OF NATURAL RESOURCE PILOTS
Francis N. Satterlee, Newsletter Editor and Public Affairs Officer
200 Patrick St., S.W.
Vienna, VA 22180
(703)560-1271

INTERNATIONAL AVIATION THEFT BUREAU
P.O. Box 3443
Frederick, MD 21701
(301)694-5444

INTERNATIONAL BIRD DOG ASSOCIATION
Phil Phillips, President
3939 San Pedro, N.E., Suite C-8
Albuquerque, NM 87110
(505)881-7555

INTERNATIONAL CESSNA 120/140 ASSOCIATION
Bill Rhoades
Box 830092
Richardson, TX 75083-0092
(612)652-2221

INTERNATIONAL CESSNA 170 ASSOCIATION
P.O. Box 1667
Lebanon, MO 65536
(417)532-4847

INTERNATIONAL CESSNA 180/185 CLUB
Charles Bombardier, President
4539 North 49th Ave.
Phoenix, AZ 85031

INTERNATIONAL CIVIL AVIATION ORGANIZATION
1000 Sherbrook St. W., Suite 327
Montreal, P.Q., Canada H3A 2R2
(514)285-8219

INTERNATIONAL COMANCHE SOCIETY, INC.
Jack A. Holaway
P.O. Box 400
Grant, NE 69140
(308)352-4275

INTERNATIONAL FLIGHT ATTENDENTS ASSOCIATION
P.O. Box 2449, Munster Str 19
D-6500, Mainz 1
West Germany
06131-234226

INTERNATIONAL FLYING FARMERS
T.W. Anderson, Executive Director
Airport Rd., P.O. Box 9124
Wichita, KS 67277

INTERNATIONAL FLYING NURSES ASSOCIATION
Dorothy Roach
322 E. Hirsch
North Lake, IL 60164

INTERNATIONAL ORGANIZATION OF AVIATION CHARACTERS
Dr. Herbert O. Fisher
Executive Vice-President
628 Mountain Rd., Smoke Rise
Kinnelon, JN 07405
(201)838-2040

INTERNATIONAL PIETNPOL ASSOCIATION
Robert L. Taylor, Editor
P.O. Box 127
Blakesburg, IA 52536
(515)938-2773

INTERNATIONAL SOCIETY OF WOMEN AIRLINE PILOTS, (ISA + 121)
Abigail F. Davis, President
P.O. Box 17156
Minneapolis, MN 55417

INTERNATIONAL TAYLORCRAFT OWNERS ASSOCIATION
Barney Bixler
12809 Greenbower Rd.
Alliance, OH 44601

THE INTERSTATE CUB
Robert L. Taylor, Editor
P.O. Box 127
Blakesburg, IA 52536
(515)938-2773

L-4 GRASSHOPPER WING
(affiliated with Cub Club)
P.O. Box 2002
Mount Pleasant, MI 48858
(517)561-2393

LAWYER-PILOTS BAR ASSOCIATION
John S. Yodice
500 E St., S.W., Suite 930
Washington, DC 20024
(202)863-1000

LIGHTER THAN AIR SOCIETY
1800 Triplett Blvd.
Akron, OH 44306
(216)325-7087 or (216)794-1321

LITTLE ROUND ENGINE FLYER
Ken Williams, Chairman
331 East Franklin St.
Portage, WI 53901

LUSCOMBE ASSOCIATION
John Bergeson
6438 West Millbrook Rd.
Remus, MI 49340
(517)561-2393

MARINE CORPS AVIATION ASSOCIATION, INC.
P.O. Box 296
Quantico, VA 22134
(703)640-6161

MAULE AIRCRAFT ASSOCIATION
Dave Neumeister, Publisher
(newsletters for AA-1, AA-5, Arctic, Arrow, Helio, Maule, Rallye, Skipper, Snow, Tomahawk, Varga, and Wren aircraft)
5630 South Washington
Lansing, MI 48911-4999
(517)882-8433

MEYERS AIRCRAFT OWNERS ASSOCIATION
Vince Vanderford, President
5852 Boque Rd.
Yuba City, CA 95991
(916)673-2724

MICHIGAN HELICOPTER ASSOCIATION
Robert J. Smith, President
P.O. Box 2613
Southfield, MI 48037
(313)728-9109

MID-ATLANTIC HELICOPTER ASSOCIATION
Mr. Jack Thompson
1611 North Kent St., Suite 910
Arlington, VA 22209

MOONEY AIRCRAFT PILOTS ASSOCIATION
314 Stardust Dr.
San Antonio, TX 78228
(512)434-5959

NATIONAL AERONAUTICS ASSOCIATION (NAA)
1763 R St., N.W.
Washington, DC 20009
(202)265-8720

NATIONAL AERONCA ASSOCIATION
Augie Wegner, Secretary/Treasurer
266 Lamp & Lantern Village
Chesterfield, MO 63017
(314)391-8999

NATIONAL AGRICULTURAL AVATION ASSOCIATION
Harold Collins, Executive Director
1005 E St., S.E.
Washington, DC 20003
(202)546-5722

NATIONAL AIRCRAFT FINANCE ASSOCIATION
Karen Griggs
500 E St., S.W., Suite 930
Washington, DC 20024
(202)554-5570

NATIONAL AIR RACING GROUP, INC.
Frank J. Ronco, Treasurer
1313 Los Arboles
Sunnyvale, CA 94087
(408)733-7967

National Air Transportation Association (NATA)
4226 King St.
Alexandria, VA 22302
(703)845-9000

National Association of Air Traffic Specialists
Bruce Henry, President
4780 Corridor Place, Suite B
Beltsville, MD 20705
(301)595-2012

National Association of Flight Examiners (NAFE)
Ohio State University Airport
P.O. Box 793
Dublin, OH 43017
(614)889-6148

National Association of Flight Instructors (NAFI)
Ohio State University Airport
P.O. Box 793
Dublin, OH 43017
(614)889-6148

National Association of Minorities in Aviation
J. Anthony Sharp, President
P.O. Box 1702
Hampton, VA 23669
(804)245-7595

National Association of Priest Pilots
Father John Hemann
510 1st Ave., N.W.
Cedar Rapids, IA 52405
(319)366-5369

National Association of State Aviation Officials (NASAO)
Robert T. Warner
Executive Vice-President
8401 Colesville Rd., Suite 505
Silver Spring, MD 20910
(301)588-0587
Fax: (301)588-1288

National Avionics Society
P.O. Box 23055
Richfield, MN 55423
(612)866-8800

National Broadcast Pilots Association
Leo Galanis Jr.
KUSA-9 TV, 1089 Bannock St.
Denver, CO 80204
(303)893-9000

National Bucker Club
Frank Price, President
Route 1, P.O. Box 419
Moody, TX 76557
(817)853-2008

National Business Aircraft Association, Inc. (NBAA)
1200 18th St., N.W., Suite 200
Washington, DC 20036
(202)783-9000

National Championship Air Races
Susan Audrain, Marketing Director
P.O. Box 1429
Reno, NV 89505
(702)826-7500

National EMS Pilots Association
P.O. Box 2354
Pearland, TX 77588
(713)997-2563

National Flight Nurses Association
Susan Baumgartner
Executive Director
P.O. Box 8222
Rapid City, SD 57709
(605)343-4464

National Flight Paramedics Association
Erroll Babineaux
P.O. Box 610487
Dallas-Fort Worth Airport
Dallas, TX 75261-0487
(817)491-1358

NATIONAL INTERCOLLEGIATE FLYING ASSOCIATION (NIFA)
4627 Ocean Blvd., No. 220
San Diego, CA 92109
(619)270-7114

NATIONAL RYAN CLUB
Bill J. Hodges, Chairman
811 Lydia
Stephenville, TX 76401
(817)968-4818

NATIONAL STINSON CLUB
Jonsey Paul
14418 Skinner Rd.
Cypress, TX 77429
(713)373-0418

NATIONAL STINSON CLUB (108 Section)
George H. Leamy, President
117 Lanford Rd.
Spartanburg, SC 29301
(803)576-9698

NATIONAL TRANSPORTATION SAFETY BOARD BAR ASSOCIATION
P.O. Box 65461
Washington, DC 20035-5461
(202)331-1955

NATIONAL WACO CLUB
Ray Brandly
700 Hill Ave.
Hamilton, OH 45015

NATIONAL WORLD WAR II GLIDER PILOTS ASSOCIATION
1500 Corinth St.
P.O. Box 15782
Dallas, TX 75215

NATIONAL 210 OWNERS ASSOCIATION
John M. Stratton, President
P.O. Box 708
La Canada Flintridge, CA 91011
(818)952-6212

NAVAL HELICOPTER ASSOCIATION
Pam Vulte
P.O. Box 460
Coronado, CO 92118
(619)437-7139

NEGRO AIRMEN INTERNATIONAL
John W. Hicks
P.O. Box 1340
Tuskegee, AL 36008

NEW ENGLAND HELICOPTER PILOTS ASSOCIATION
John Anderson, President
P.O. Box 88
Bedford, MA 01730

NINETY-NINES, INC.
(international organization of women pilots)
Loretta Gragg, Executive Director
Will Rogers World Airport
P.O. Box 59965
Oklahoma City, OK 73159
(405)685-7969

NORTH AMERICAN TRAINER ASSOCIATION (T6, T-28, NA64, NA50)
Stoney and Kathy Stonich
2285 Oakvale Dr.
Shingle Springs, CA 95682
(916)677-2456

NORTHEAST STINSON FLYING CLUB
Dick Bourque, Founder
8 Grimes Brook Rd.
Simsbury, CT 06070
(203)658-1566

ORDER OF DAEDALIANS
(Fraternal Order of Military Pilots)
Building 1635
Kelly Air Force Base, TX 78241
(512)924-9485

ORGANIZATION OF FLYING ADJUSTERS (OFA)
B.R. Mitchell, President
P.O. Box 75637
Oklahoma City, OK 73147
(405)942-7171

OX-5 AVIATION PIONEERS
207 Dormont Village
2961 West Liberty Ave.
Pittsburgh, PA 15216
(412)341-5650

P-40 WARHAWK PILOTS ASSOCIATION
Dr. Herbert O. Fisher
Director of Public Relations and Press
628 Mountain Rd., Smoke Rise
Kinnelon, NJ 07405
(201)838-2040

P-47 THUNDERBOLT PILOTS ASSOCIATION
Dr. Herbert O. Fisher
Director of Public Relations and Press
628 Mountain Rd., Smoke Rise
Kinnelon, NJ 07405
(201)838-2040

P-47 ALUMNI ASSOCIATION
Dr. Herbert O. Fisher
Director of Communications
and Honorary Trustee
628 Mountain Rd., Smoke Rise
Kinnelon, NJ 07405
(201)838-2040

P-51 MUSTANG PILOTS ASSOCIATION
Dr. Herbert O. Fisher
Director of Public Relations
628 Mountain Rd., Smoke Rise
Kinnelon, NJ 07405
(201)838-2040

PILOTS INTERNATIONAL
ASSOCIATION, INC.
4000 Olson Memorial Hwy.
Minneapolis, MN 55422-5397
(612)588-5175

POPULAR ROTORCRAFT
ASSOCIATION, INC.
P.O. Box 8756
Clinton, LA 70722
(504)683-3545

PORTERFIELD AIRPLANE CLUB
Chuck Lebrecht
1019 Hickory Rd.
Ocala, FL 32672
(904)687-4859

POWDERED ULTRALIGHT
MANUFACTURERS ASSOCIATION (PUMA)
P.O. Box 380220
Duncansville, TX 75138
(214)298-7033

PROFESSIONAL AEROMEDICAL TRANSPORT
ASSOCIATION
Dr. Wesley Bare, President
P.O. Box 579
Moorestown, NJ 08057
(800)257-8180 or (609)234-0330

PROFESSIONAL AVIATION MAINTENANCE
ASSOCIATION, INC.
David Wadsworth, Executive Director
500 N.W. Plaza, Suite 912
St. Ann, MO 63074
(314)739-2850

PROFESSIONAL RACE PILOTS
ASSOCIATION (PRPA)
4895 Texas Ave.
P.O. Box 60084
Reno, NV 89506
(702)322-1421

REAL ESTATE AVIATION CHAPTER OF THE
FARM AND LAND INSTITUTE
National Association of Realtors
Warren J. Haeger
c/o Indust-Realty, Incorporated
5440 Saint Charles Rd.
Berkeley, IL 60163
(312)547-7100

REARWIN CLUB
Robert L. Taylor, Editor
P.O. Box 127
Blakesburg, IA 52536
(515)938-2773

REGIONAL AIRLINE ASSOCIATION
1101 Connecticut Ave., N.W.
Suite 700
Washington, DC 20036
(202)857-1170

REPLICA FIGHTERS ASSOCIATION
Frank G. Weatherly, President
2789 Mohawk Lane
Rochester, MI 48064
(313)651-7008

SEABEE CLUB INTERNATIONAL
6761 N.W. 32nd Ave.
Fort Lauderdale, FL 33309
(305)979-5470

SEAPLANE PILOTS ASSOCIATION
Robert A. Richardson,
Executive Director
421 Aviation Way
Frederick, MD 21701
(301)695-2000

SHORT WING PIPER CLUB, INC.
Lonnie McLaughlin
32 West End Ave.
Brentwood, NY 11717
(516)273-5072

SILVER WINGS FRATERNITY
Russ Brinkley, President
Box 11970
Harrisburg, PA 17108
(717)232-9525

SOARING SOCIETY OF AMERICA, INC.
Larry Sanderson, Executive Director
P.O. Box E
Hobbs, NM 88241
(505)392-1177

SOCIETY OF AUTOMOTIVE ENGINEERS, INC.
Dave Bentley
Aerospace Program Manager
400 Commonwealth Dr.
Warrendale, PA 15096
(412)776-4841

SOCIETY OF EXPERIMENTAL TEST PILOTS
Thomas H. Smith, Executive Director
44814 North Elm St.
Lancaster, CA 93534
(805)942-9574

SOCIETY OF FLIGHT TEST ENGINEERS, INC.
P.O. Box 4047
Lancaster, CA 93539
(805)948-3067

SOUTHWEST STINSON CLUB
Dick Goerges, President
3619 Nortree St.
San Jose, CA 95148
(408)274-9179

SPARTAN SCHOOL OF AERONAUTICS ALUMNI ASSOCIATION
Ted Knotts
8820 East Pine St.
Tulsa, OK 74115
(918)836-6886

STAGGERWING CLUB
Jim Gorman, President
1885 Millsboro Rd.
Mansfield, OH 44906
(419)529-3822

STEARMAN RESTORERS ASSOCIATION
Tom Lowe
823 Kingston Lane
Crystal Lake, IL 60014
(815)459-6873

SUPER CUB PILOTS ASSOCIATION
Jim Richmond, Founder-Director
P.O. Box 9823
Yakima, WA 98909
(509)248-9491

SWIFT ASSOCIATION
McMinn County Airport
Box 644
Athens, TN 37303
(615)745-9547

TAMPA BAY HELICOPTER COUNCIL
Stephanie V. Slavin
7507 South Tamiami Trail, Suite 176
Sarasota, FL 34231
(813)921-4941

Taylorcraft Owners Club
Bruce M. Bixler, President
12809 Greenbower Rd.
Alliance, OH 44601
(216)823-9748

Texas Pilots Association
3426 Hillsdale Lane
Garland, TX 75046-5546
(214)487-7032

301st Veterans Association
Billy S. McCarty, Secretary
P.O. Box 47843
San Antonio, TX 78265-8843
(512)654-3729

Tomahawk Owners and Operators of North America (TOONA)
Calypso Airways Terminal
Easton Municipal Airport
Easton, MD 21601
(301)822-8854

Travel Air Club
Robert L. Taylor, Editor
P.O. Box 127
Blakesburg, IA 52536
(515)938-2773

Twin Bonanza Association
Richard I. Ward
19684 Lakeshore Dr.
Three Rivers, MI 49093
(616)279-2540

The Twirly Birds
John Slattery
P.O. Box 18029
Oxon Hill, MD 20745
(301)567-4407

United Flying Octogenarians
A.R. Boileau
3531 Larga Circle
San Diego, CA 92110

United States Air Traffic Controllers
210 Seventh St. S.E., Suite C-26
Washington, D.C. 20003

United States Hang Gliding Association, Inc.
(NAA affiliate)
P.O. Box 500
Pearblossom, CA 93553
(805)944-5333

United States Parachute Association
1440 Duke St.
Alexandria, VA 22314
(703)836-3495

United States Pilots Association
483 South Kirkwood Rd., Suite 10
St. Louis, MO 63122
(314)843-2766

University Aviation Association
Gary W. Kitely, Executive Director
3410 Skyway Dr.
Opelika, AL 36801
(205)826-2308

Vietnam Helicopter Pilots Association
P.O. Box 9592
Wichita, KS 67277
(316)946-4047

Vintage Sailplane Association
Jan Scott, Secretary
Route 1, P.O. Box 239
Lovettsvile, VA 22080
(703)822-5504

Waco Historical Society
R.E. Hoefflin, Treasurer
1013 Westgate Rd.
Troy, OH 45373
(513)335-2621

Warbirds of America
Donald C. Davidson, President
Wittman Field
Oshkosh, WI 54903-3086
(414)426-4800

Warbirds Worldwide, Ltd.
Paul A. Coggan, Director
19 Highcliffe Ave.
Shirebrook
Mansfield, England NG20 8NB

WEST COAST CESSNA 120/140 CLUB
Elsie Thompson
P.O. Box 727
Roseburg, OR 97470
(503)672-5046

WESTERN BONANZA BEECH OWNERS SOCIETY
Alden C. Barrios, President
1436 Muirlands Dr.
La Jolla, CA 92037
(619)459-5901

WHIRLY-GIRLS, INC.
(international organization of women helicopter pilots)
1619 Duke St.
Alexandria, VA 22314-3406
(703)683-4646

WORLD WAR I AEROPLANES, INC.
Leonard E. Opdyke
Director-Publisher
15 Crescent Rd.
Poughkeepsie, NY 12601
(914)473-3679

WORLDWIDE AVIATORS FLYING PROFESSIONAL AID
2950 Mission Dr., St. 16
P.O. Box 1115
Solvang, CA 93463
(805)688-3458

ZENAIR ASSOCIATION
6438 West Millbrook Rd.
Remus, MI 49340
(517)561-2393

Appendix Aviation Statistics

Active General Aviation Aircraft (as of Dec. 31, 1986)

Total	220,044
Fixed-wing	206,090
Single-engine, piston	171,777
Multiengine, piston	23,721
Other	148
Turboprop	5,964
Turbofan/turbojet	4,480
Rotary wing	6,943
Piston	2,921
Turbine	4,022
Other	7,010

Air Carrier Aircraft (as of Dec. 31, 1986)

Total	4,909
Fixed-wing	4,907
Piston	420
Turboprop	1,204
Turbofan/turbojet	3,283
Rotary wing	2

Active Pilots (as of Dec. 31, 1987)

Total	699,653
Student	146,016

Private	300,949
Commercial	143,645
Airline Transport	91,287
Helicopter, only	8,702
Glider, only	7,901
Lighter-than-air	1,153
Instrument-rated pilots	266,122
Flight instructors	60,316

Civil Aircraft Activity (total hours flown in 1986)

Total	48.1 million
General aviation	34.4 million
Air carrier	13.7 million

Landing Facilities (as of Dec. 31, 1987)

Total	17,015
Airports	12,907
Heliports	3,653
Stolports	67
Seaplane bases	388
Publicly owned	4,984
Privately owned	12,031
Public use	5,723
Private use	11,292

Busiest U.S. Airports (aircraft operations in fiscal 1987)

1. Atlanta Hartsfield International	801,833
2. Chicago O'Hare International	796,609
3. Los Angeles International	655,189
4. Dallas/Ft. Worth Regional	609,300
5. Santa Ana	526,798
6. Denver Stapleton International	521,608
7. Van Nuys	492,936
8. San Francisco	451,132
9. Long Beach	438,496
10. Boston Logan	435,923
11. Phoenix Sky Harbor International	435,836
12. St. Louis International	426,828
13. Philadelphia International	412,083
14. Detroit Metropolitan Wayne County	411,628
15. Oakland International	397,658
16. Honolulu	389,035
17. Las Vegas McCarran	388,962
18. Pontiac	386,292
19. Memphis International	384,049
20. Minneapolis St. Paul International	383,420

Index

D

E

F